Mechanical PE Sample Examination

Second Edition

Michael R. Lindeburg, PE

Professional Publications, Inc. • Belmont, California

Benefit by Registering This Book with PPI

- Get book updates and corrections
- Hear the latest exam news
- Obtain exclusive exam tips and strategies
- Receive special discounts

Register your book at **www.ppi2pass.com/register**.

Report Errors and View Corrections for This Book

PPI is grateful to every reader who notifies us of a possible error. Your feedback allows us to improve the quality and accuracy of our products. You can report errata and view corrections at **www.ppi2pass.com/errata**.

MECHANICAL PE SAMPLE EXAMINATION
Second Edition

Current printing of this edition: 3

Printing History

edition number	printing number	update
2	1	New edition. Major update. Copyright change.
2	2	Minor corrections.
2	3	Minor corrections.

Printed in the United States of America

PPI
1250 Fifth Avenue, Belmont, CA 94002
(650) 593-9119
www.ppi2pass.com

ISBN: 978-1-59126-160-5

Library of Congress Control Number: 2009920655

Table of Contents

Preface and Acknowledgments

This second edition of *Mechanical PE Sample Examination* has been substantially updated to reflect the changes in the breakdown of subject areas covered by the mechanical PE exam that went into effect in October 2008. Where NCEES has increased or decreased the percentage of exam coverage of a particular subject area, we have added or subtracted problems from this sample exam accordingly.

This book accurately reflects the NCEES breadth-and-depth exam format, which gives you the option of selecting from among three afternoon specialty subject areas: HVAC and refrigeration, mechanical systems and materials, and thermal and fluids systems. In a sense, this book is actually three sample exams in one.

The breadth-and-depth format of the PE exam presents several challenges to examinees. The breadth of the subject matter covered by the exam requires that you have a firm grasp of mechanical engineering fundamentals. These rudiments are generally covered in an undergraduate curriculum. You still need to know how to select a pump, size a shaft, and humidify a room. However, depth problems enable NCEES to focus on, target in, and drill down to some very specific knowledge bases. The questions within each area of afternoon emphasis require knowledge gained only through experience—a true test of your worthiness of licensure. Often, problems testing proficiency in an area can be worded as simple definition questions. For example, if you don't recognize the application of a jockey pump, you probably aren't ready to design fire protection sprinkler systems.

Needless to say, the number of problems in this sample exam for the various exam topics is consistent with the current NCEES subject breakdown. The solutions in this sample exam are also consistent in nomenclature and style with the *Mechanical Engineering Reference Manual*. Hopefully, you have already made that book part of your exam-day arsenal.

In comparison to the FE (Fundamentals of Engineering) exam, which is essentially all in SI (metric) units, the mechanical PE exam continues to feature English units predominantly. However, there is a trend toward more use of SI units, and accordingly, metric units are sometimes used in this sample exam. Some use of metric units is inevitable, anyway, particularly in the areas of generated power, electrical heating, and chemical concentrations and quantities.

Most of the content of this book results directly from the work of a number of excellent engineers and professors. These include Daniel C. Deckler, PhD, PE (writing in the subject area of thermal and fluids systems), Keith Elder, PE (writing in the subject area of HVAC and refrigeration), and John M. Iaconis, PE (writing in the subject area of mechanical systems and materials). The primary content of most of the problems originated with these contributors. Material was technically reviewed for each of these subject areas by Arne Lund, Matthew H. Gordon, PhD, PE, and Steven J. Murray, PhD, PE, respectively. I am grateful for their efforts.

For this second edition, new problems have been added by David A. Bostain, PE, David T. Corbin, PE, Mahesh K. Kamatala, PE, Zane Pucylowski, PE, and me.

Editing, typesetting, illustrating, and proofreading of this book continues to follow PPI's strict style guides for engineering publications. I know you will benefit greatly from the flow, completeness, and appearance of the problem statements and their solutions.

I appreciate the significant contributions to this book by my editors and project managers, Sean Sullivan, John Boykin, and Scott Marley, compositors Kate Hayes and Miriam Hanes, illustrator Jamie Gibson, compositor, illustrator, and cover artist Amy Schwertman, and staff engineer Chuck Simchick. The editorial and production staffs were managed by Sarah Hubbard and Cathy Schrott, respectively. Without their efforts, this edition would read like a bunch of scribbles.

As in all of my publications, I invite your comments. If you disagree with a solution, or if you think there is a better way to do something, please let me know. You can submit errata online at the PPI website at **www.ppi2pass.com/errata**.

Best wishes in your exam and subsequent career.

Michael R. Lindeburg, PE

Codes Used in This Book

PPI lists on its website the dates and editions of the codes, standards, and regulations on which NCEES has announced it will base the PE exam. These NCEES postings do not contain any announcements specifically about the mechanical engineering exam. (ASME and ASTM standards and codes are listed, but there is no mention of specific dates or editions.) The conclusion that you and I must reach from such an omission is that the exam is not particularly sensitive to changes in codes, standards, regulations, or announcements in the Federal Register. Nevertheless, the credibility of your review demands that you study from and prepare to use only the last edition of a code issued before the year of your exam.

This absence of specificity also implies that having a specific code reference in your possession on the day of the exam is not necessary. For example, having the entirety of the NFPA standards by your side is not necessary. (Remember that jockey pump mentioned in the Preface and Acknowledgments?) You won't be expected to extract and regurgitate specific data from a code or standard.

NCEES reports that it has taken the approach of providing in the exam any code sections that it supposes you would not and should not have memorized. Thus, if NCEES wants to test your ability to navigate through and use a code section, it should provide that code section as part of the problem statement. On the other hand, if NCEES supposes that "any engineer worth his or her salt would already know this," you'll have to answer the question without the benefit of a reference.

What that means, basically, is that preparing for code questions in areas in which you do not work is so difficult as to be essentially impossible. Thankfully, the number of code questions on the exam is small.

Introduction

ABOUT THE PE EXAM

The *Mechanical PE Sample Examination* provides the opportunity to practice taking an eight-hour test similar in content and format to the Principles and Practice of Engineering (PE) examination in mechanical engineering. The mechanical PE examination is an eight-hour exam divided into a morning session and an afternoon session. The morning session is known as the "breadth" exam, and the afternoon is known as the "depth" exam. This book contains a sample breadth module and three sample depth modules—one for each subdiscipline the NCEES tests.

In the four-hour morning session, the examinee is asked to solve 40 problems from five major mechanical engineering subdisciplines: basic engineering practice (approximately 30% of the exam problems); mechanical systems and materials (20%); hydraulics and fluids (17%); energy/power systems (15%); and HVAC and refrigeration (18%). Morning session problems are general in nature and wide-ranging in scope.

The four-hour afternoon session allows the examinee to select a depth exam module from one of three subdisciplines (HVAC and refrigeration, mechanical systems and materials, and thermal and fluids systems). Each depth module is made up of 40 problems. Afternoon session problems require more specialized knowledge than those in the morning session.

All problems, from both the morning and afternoon sessions, are multiple choice. They include a problem statement with all required defining information, followed by four logical choices. Only one of the four options is correct. The problems are completely independent of each other, so an incorrect choice on one problem will not carry over to subsequent problems.

This book is written in the multiple-choice exam format instituted by the NCEES. It covers all the same topic areas that appear on the exam, as provided by the NCEES.

Topics and the approximate distribution of problems on the morning session of the mechanical PE exam are as follows.

Basic Engineering Practice:
approximately 30% of exam problems

Mechanical Systems and Materials:
approximately 20% of exam problems

- Principles: 13%
- Applications: 7%

Hydraulics and Fluids:
approximately 17% of exam problems

- Principles: 7%
- Applications: 10%

Energy/Power Systems:
approximately 15% of exam problems

- Principles: 7%
- Applications: 8%

HVAC and Refrigeration:
approximately 18% of exam problems

- Principles: 10%
- Applications: 8%

Topics and the approximate distribution of problems on the afternoon sessions of the mechanical PE exam are as follows.

HVAC and Refrigeration Module

- Principles: approximately 55% of exam problems
- Applications: approximately 45% of exam problems

Mechanical Systems and Materials Module

- Principles: approximately 60% of exam problems
- Applications: approximately 40% of exam problems

Thermal and Fluids Systems Module

- Principles: approximately 45% of exam problems
- Applications: approximately 55% of exam problems

For further information and tips on how to prepare for the mechanical PE exam, consult the *Mechanical Engineering Reference Manual* or visit PPI's website, **www.ppi2pass.com/mefaq**.

HOW TO USE THIS BOOK

This book is a sample exam—the main issue is not *how* you use it, but *when* you use it. It was not intended to be a diagnostic tool to guide your preparation. Rather, its value is in giving you an opportunity to bring together all of your knowledge and to practice your test-taking skills. The three most important skills are (1) selection of the right subjects to study, (2) organization of your references and other resources, and (3) time management. Take this sample exam within a few weeks of your actual exam. That's the only time that you will be able to focus on test-taking skills without the distraction of rusty recall.

Do not read the questions ahead of time, and do not look at the answers until you've finished. Prepare for the sample exam as you would prepare for the actual exam. Assemble your reference materials. Check with your state's board of engineering registration for any restrictions on what materials you can bring to the exam. (The PPI website, **www.ppi2pass.com/mefaq**, has a listing of state boards.) Read the sample exam instructions (which simulate the ones you'll receive from your exam proctor), set a timer for four hours, and take the breadth module. After a one-hour break, turn to the depth module you will select during the actual exam, set the timer, and complete the simulated afternoon session. Then, check your answers.

The problems in this book were written to emphasize the breadth of the mechanical engineering field. Some may seem easy and some hard. If you are unable to answer a problem, you should review that topic area.

This book assumes that the breadth module of the PE exam will be more academic and traditional in nature, and that the depth modules will require practical, non-numerical knowledge of the type that comes from experience.

The problems are generally similar to each other in difficulty, yet a few somewhat easier problems have been included to expose you to less frequently examined topics.

After taking the sample exam, review your areas of weakness and then take the exam again, but substitute a different depth module. Check your answers, and repeat the process for each of the depth areas. Evaluate your strengths and weaknesses, and select additional texts to supplement your weak areas. Check the PPI website for the latest in exam preparation materials at **www.ppi2pass.com**.

The keys to success on the exam are to know the basics and to practice solving as many problems as possible. This book will assist you with both objectives.

Pre-Test

You can use the following incomplete table to judge your preparedness. You should be able to fill in all of the missing information. (The completed table appears on the back of this page.) If you are ready for the sample exam (and, hence, for the actual exam), you will recognize them all, get most correct, and when you see the answers to the ones you missed, you'll say, "Ahh, yes." If you have to scratch your head with too many of these, then you haven't exposed yourself to enough of the subjects that are on the exam.

description	**value or formula**	**units**
acceleration of gravity, g		in/sec^2
gravitational constant, g_c	32.2	
formula for the area of a circle		ft^2
	1545	ft-lbf/lbmol-°R
	+ 460	°
density of water, approximate		lbm/ft^3
density of air, approximate	0.075	
specific gas constant for air	53.3	
	1.0	Btu/lbm-°R
foot-pounds per second in a horsepower		ft-lbf/hp-sec
pressure, p, in fluid with density, ρ, in lbm/ft^3, at depth, h		lbf/ft^2
cancellation and simplification of the units (A = amps; rad = radians)	A-sec^4/sec^5-rad	
specific heat of air, constant pressure		Btu/lbm-°R
common units of entropy of steam	–	
	$bh^3/12$	cm^4
molecular weight of oxygen gas		lbm/lbmol
what you add to convert $\Delta T_{°F}$ to $\Delta T_{°R}$		°
	Q/A	ft/sec
inside surface area of a hollow cylinder with length L and diameters d_i and d_o		ft^2
the value that NPSHA must be larger than		ft
	849	lbm/ft^3
what you have to multiply density in lbm/ft^3 by to get specific weight in lbf/ft^3		lbf/lbm
power dissipated by a device drawing I amps when connected to a battery of V volts		W
linear coefficient of thermal expansion for steel	6.5×10^{-6}	
formula converting degrees centigrade to degrees Celsius		
the primary SI units constituting a newton of force		N
the difference between psig and psia at sea level		psi
the volume of a mole of an ideal gas		ft^3
	1.713×10^{-9}	Btu/ft^2-hr-°R^4
universal gas constant	8314	
	2.31	ft/psi
shear modulus of steel		psi
conversion from rpm to rad/sec		rad-min/rev-sec
Joule's constant	778	

PRE-TEST ANSWER KEY

description	value or formula	units
acceleration of gravity, g	386	in/sec^2
gravitational constant, g_c	32.2	ft-lbm/lbf-sec^2
formula for the area of a circle	πr^2 or $(\pi/4)d^2$	ft^2
universal gas constant in customary U.S. units	1545	ft-lbf/lbmol-°R
what you add to $T_{\circ F}$ to obtain $T_{\circ R}$ (absolute temperature)	+ 460	°
density of water, approximate	62.4	lbm/ft^3
density of air, approximate	0.075	lbm/ft^3
specific gas constant for air	53.3	ft-lbf/lbm-°R
specific heat of water	1.0	Btu/lbm-°R
foot-pounds per second in a horsepower	550	ft-lbf/hp-sec
pressure, p, in fluid with density, ρ, in lbm/ft^3, at depth, h	$p = \gamma h = \rho g h / g_c$	lbf/ft^2
cancellation and simplification of the units (A = amps; rad = radians)	A-sec^4/sec^5-rad	W (watts)
specific heat of air, constant pressure	0.241	Btu/lbm-°R
common units of entropy of steam	–	Btu/lbm-°R
centroidal moment of inertia of a rectangle	$bh^3/12$	cm^4
molecular weight of oxygen gas	32	lbm/lbmol
what you add to convert $\Delta T_{\circ F}$ to $\Delta T_{\circ R}$	0	°
velocity of flow	Q/A	ft/sec
inside surface area of a hollow cylinder with length L and diameters d_i and d_o	$\pi d_i L$	ft^2
the value that NPSHA must be larger than	h_v (vapor head)	ft
density of mercury	849	lbm/ft^3
what you have to multiply density in lbm/ft^3 by to get specific weight in lbf/ft^3	g/g_c (numerically, 32.2/32.2 or 1.0)	lbf/lbm
power dissipated by a device drawing I amps when connected to a battery of V volts	IV	W
linear coefficient of thermal expansion for steel	6.5×10^{-6}	ft/ft-°F or ft/ft-°R
formula converting degrees centigrade to degrees Celsius	°centigrade = °Celsius	The centigrade scale is obsolete.
the primary SI units constituting a newton of force	kg-m/s^2	N
the difference between psig and psia at sea level	14.7 psia (atmospheric pressure)	psi
the volume of a mole of an ideal gas	359 or 360	ft^3
Stefan-Boltzmann constant	1.713×10^{-9}	Btu/ft^2-hr-°R^4
universal gas constant	8314	J/kmol·K
conversion from psi to height of water	2.31	ft/psi
shear modulus of steel	11.5×10^6	psi
conversion from rpm to rad/sec	$2\pi/60$	rad-min/rev-sec
Joule's constant	778	ft-lbf/Btu

Morning Session
Instructions

In accordance with the rules established by your state, you may use textbooks, handbooks, bound reference materials, and any approved battery- or solar-powered, silent calculator to work this examination. However, no blank papers, writing tablets, unbound scratch paper, or loose notes are permitted. Sufficient room for scratch work is provided in the Examination Booklet.

You are not permitted to share or exchange materials with other examinees. However, the books and other resources used in this morning session may be changed prior to the afternoon session.

You will have four hours in which to work this session of the examination. Your score will be determined by the number of questions that you answer correctly. There is a total of 40 questions. All 40 questions must be worked correctly in order to receive full credit on the exam. There are no optional questions. Each question is worth one point. The maximum possible score for this section of the examination is 40 points.

Partial credit is not available. No credit will be given for methodology, assumptions, or work written in your Examination Booklet.

Record all of your answers on the Answer Sheet. No credit will be given for answers marked in the Examination Booklet. Mark your answers with the pencil provided to you. Marks must be dark and must completely fill the bubbles. Record only one answer per question. If you mark more than one answer, you will not receive credit for the question. If you change an answer, be sure the old bubble is erased completely; incomplete erasures may be misinterpreted as answers.

If you finish early, check your work and make sure that you have followed all instructions. After checking your answers, you may turn in your Examination Booklet and Answer Sheet and leave the examination room. Once you leave, you will not be permitted to return to work or change your answers.

When permission has been given by your proctor, break the seal on the Examination Booklet. Check that all pages are present and legible. If any part of your Examination Booklet is missing, your proctor will issue you a new Booklet.

Do not work any questions from the Afternoon Session during the first four hours of this exam.

WAIT FOR PERMISSION TO BEGIN

Name: ______________________________
Last First Middle Initial

Examinee number: ______________

Examination Booklet number: ______________

Principles and Practice of Engineering Examination

Morning Session Sample Examination

Morning Session

1.	A	B	C	D	11.	A	B	C	D	21.	A	B	C	D	31.	A	B	C	D
2.	A	B	C	D	12.	A	B	C	D	22.	A	B	C	D	32.	A	B	C	D
3.	A	B	C	D	13.	A	B	C	D	23.	A	B	C	D	33.	A	B	C	D
4.	A	B	C	D	14.	A	B	C	D	24.	A	B	C	D	34.	A	B	C	D
5.	A	B	C	D	15.	A	B	C	D	25.	A	B	C	D	35.	A	B	C	D
6.	A	B	C	D	16.	A	B	C	D	26.	A	B	C	D	36.	A	B	C	D
7.	A	B	C	D	17.	A	B	C	D	27.	A	B	C	D	37.	A	B	C	D
8.	A	B	C	D	18.	A	B	C	D	28.	A	B	C	D	38.	A	B	C	D
9.	A	B	C	D	19.	A	B	C	D	29.	A	B	C	D	39.	A	B	C	D
10.	A	B	C	D	20.	A	B	C	D	30.	A	B	C	D	40.	A	B	C	D

Morning Session

1. Based on chemical resistance alone, which of the following pipe materials would provide the best performance?

(A) polyvinyl chloride (PVC)
(B) polystyrene
(C) polypropylene
(D) acrylonitrile butadiene styrene (ABS)

2. A portion of a pin-connected truss is shown. During an earthquake, pin E experiences a vertical load that alternates between 5 kips upward and 10 kips downward.

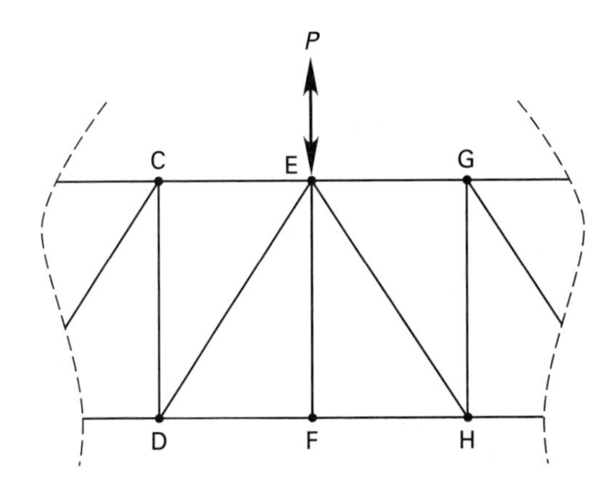

The loading in member EF will

(A) alternate between tension and compression
(B) always be compression
(C) always be tension
(D) always be zero

3. The water level of a 15 psig pressurized tank is 8 ft below the water level of an open tank. A pump 15 ft below the water level of the open tank delivers water through schedule-40 steel pipe to the pressurized tank. The losses due to both pipe friction and fittings are 12 ft. The pump must have a total dynamic head most nearly of

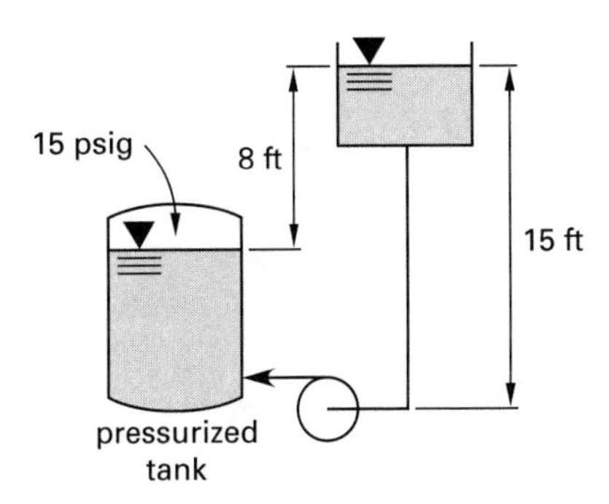

(A) 19 ft water
(B) 27 ft water
(C) 39 ft water
(D) 54 ft water

4. The illustration shown here is an example of a

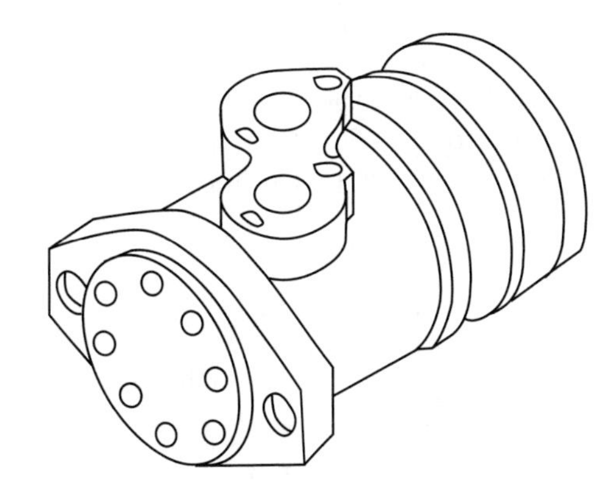

(A) cavalier view
(B) isometric view
(C) principal view
(D) sectional view

5. When running at full load, a chiller has the capacity to cool 640 gpm of water from 55°F to 43°F. If the rated coefficient of performance (COP) is 4.4, the total compressor heat that must be rejected to the cooling tower is most nearly

(A) 21 tons
(B) 73 tons
(C) 320 tons
(D) 390 tons

6. At a remote jobsite, a 12 V direct current inverter provides 120 V alternating current power from a battery to run a 0.6 kW orbital reciprocating saw. The inverter has a resistance of 0.16 Ω and an average efficiency of 68%. What is the approximate effective current flowing to the inverter?

(A) 5.0 A
(B) 7.3 A
(C) 34 A
(D) 74 A

7. A hydraulic reservoir with a diameter of 30 in is open to the atmosphere and has a 1/8 in diameter relief check valve feeding a return line. The flow coefficient for the check valve is 0.53, and the valve's cracking pressure is 5 psi. The reservoir contains 24 in of hydraulic oil having a specific gravity of 0.90 at an operating temperature of 100°F. When the oil pressure is 5.775 psig, what is the approximate flow rate of the hydraulic oil into the reservoir from the return line?

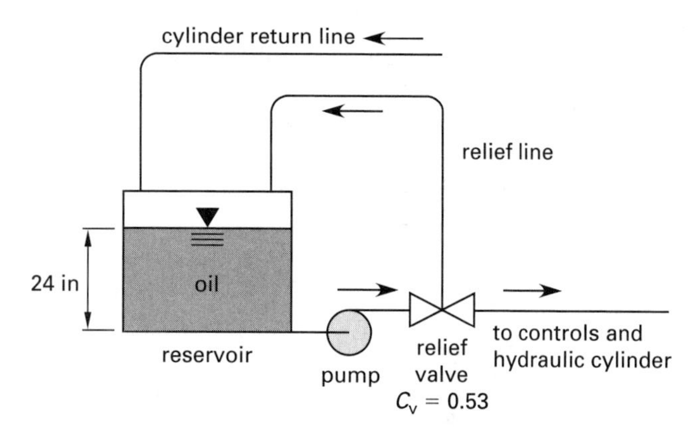

(A) 0.41 gpm
(B) 0.53 gpm
(C) 1.3 gpm
(D) 1.4 gpm

8. An aerospace application requires the adhesive bonding of two sheets of titanium skin to form an assembly that can resist relatively high flexural loading. Budgetary restrictions are not a factor. Which of the following types of joints is most appropriate?

(A) double butt lap joint
(B) double scarf lap joint
(C) joggle lap joint
(D) tapered single lap joint

9. The kinematic viscosity of an SAE 10W-30 engine oil with a specific gravity of 0.88 is reported as 110 centistokes at 37°C. Expressed in Saybolt universal seconds, the viscosity is most nearly

(A) 0.9 SUS
(B) 49 SUS
(C) 500 SUS
(D) 32,000 SUS

10. An environmental criterion for a museum requires 8000 ft^3/min supply air to be delivered to a gallery space at 60°F db and 51°F dew point. If the condition of the air leaving the cooling coil is 53°F db and 52°F wb, the reheat load is most nearly

(A) 17,000 Btu/hr
(B) 60,000 Btu/hr
(C) 69,000 Btu/hr
(D) 78,000 Btu/hr

11. Dry air flows into a long, insulated channel that contains a pool of water. Liquid water flows in at point 2 at the same rate as the evaporation rate to maintain a constant quantity of liquid. The air flows in through the inlet at point 1 at a rate of 12 kg/s at a temperature of 25°C and pressure of 100 kPa. If the air exits at point 3 saturated at 40°C, what is most nearly the channel water's rate of evaporation?

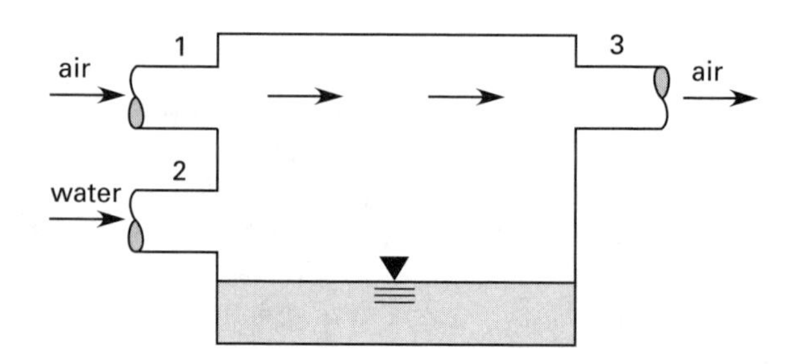

(A) 0.04 kg/s
(B) 0.6 kg/s
(C) 10 kg/s
(D) 100 kg/s

12. A 10 m × 6 m × 3 m room contains air at 33°C and 50 kPa at a relative humidity of 40%. What is the partial pressure of the water vapor in the room?

(A) 2.0 kPa
(B) 4.0 kPa
(C) 45 kPa
(D) 55 kPa

13. An air conditioning system takes in outdoor air at 5°C and 20% relative humidity at a steady rate of 1.0 m^3/s. Once inside the system, the air is heated to 21°C and then humidified with hot steam to a final temperature of 25°C and 50% relative humidity. If the humidification process takes place at 100 kPa, what is the required mass flow rate of the hot steam?

(A) 0.0014 kg/min
(B) 0.11 kg/min
(C) 0.67 kg/min
(D) 0.73 kg/min

14. In a Brayton refrigeration cycle, 2500 cfm of air enter the compressor at 14.7 psia and 25°F. The upper pressure in the cycle is 44.1 psia, and the turbine inlet temperature is 90°F. The turbine efficiency is 85%, and the compressor efficiency is 80%. The refrigeration capacity of the system is most nearly

(A) 15 tons
(B) 20 tons
(C) 31 tons
(D) 39 tons

15. A 20 ft length of 2½ in nominal diameter schedule-40 steel pipe full of water is simply supported. What is most nearly the maximum pipe deflection?

(A) 0.18 in
(B) 0.45 in
(C) 0.50 in
(D) 0.61 in

16. A fluid flows through a tube at a rate of 3×10^{-5} ft^3/sec. The tube is 3.4 ft long and has an internal diameter of 0.018 in. The flow is viscous with a viscosity of 26.37×10^{-6} $lbf\text{-}sec/ft^2$, incompressible, steady, and laminar. The predicted pressure drop across the length of the tube is most nearly

(A) 11,000 lbf/ft^2
(B) 15,000 lbf/ft^2
(C) 22,000 lbf/ft^2
(D) 26,000 lbf/ft^2

17. As a project management tool, a graphic record is made of the daily production rate of a construction crew, as shown.

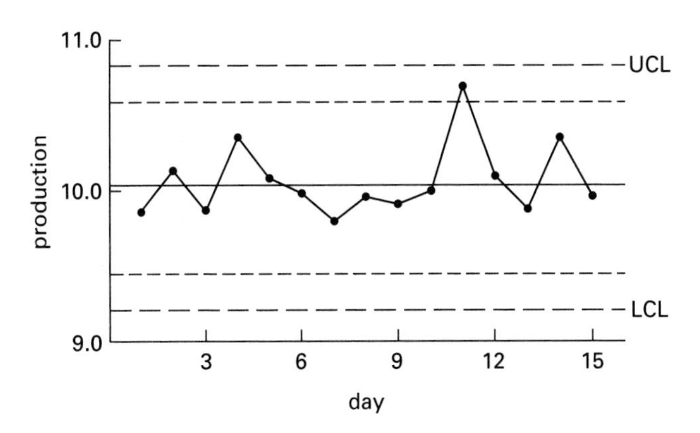

This type of monitoring record is known as a

(A) Gantt chart
(B) moving average chart
(C) *p*-chart
(D) Shewhart chart

18. A construction company purchases a 40 ton Volvo dump loader for $380,000. The company depreciates the dump loader over seven years using the MACRS method. A salvage value of $60,000 is expected at the end of seven years. The company is in the 36% income tax bracket and uses a 10% effective annual interest rate for all of its economic comparisons. After three years of operation, what after-tax depreciation recovery has been accumulated toward a replacement dump loader?

(A) $46,000
(B) $60,000
(C) $84,000
(D) $170,000

19. A welded circular tube with a 0.03 in wall thickness and a 0.50 in outside diameter experiences periodic torque varying between −3 in-lbf and 30 in-lbf. The tube is steel with a 42,000 lbf/in^2 tensile yield strength and a 72,000 lbf/in^2 tensile ultimate strength. The tube endurance strength is 24,000 lbf/in^2, and no stress concentrations exist. What is most nearly the factor of safety for infinite life fatigue loading?

(A) 1.6
(B) 3.2
(C) 4.7
(D) 6.5

20. A cantilevered flat steel spring supports a tip load that periodically varies between 0 lbf and 1 lbf. The spring is 0.75 in wide and 4.0 in long. Tensile yield strength is 75,000 lbf/in^2, tensile ultimate strength is 100,000 lbf/in^2, and endurance strength after accounting for various derating factors is 40,000 lbf/in^2. The design factor of safety is 3. The minimum required spring thickness for infinite life is most nearly

(A) 0.029 in
(B) 0.035 in
(C) 0.041 in
(D) 0.058 in

21. The end of a circular steel tube is welded to a steel plate. The full-perimeter external fillet weld has a $^3/_{16}$ in throat, and allowable stress at the throat is 10,400 lbf/in^2. The tube outside diameter is 3 in. The tube and plate thicknesses are $^1/_2$ in. A load is applied 4 in from the plate surface, parallel to the plate, perpendicular to the tube axis, and at the centroid of the tube section. Assume the stresses are uniform across the weld. What is most nearly the maximum supported load?

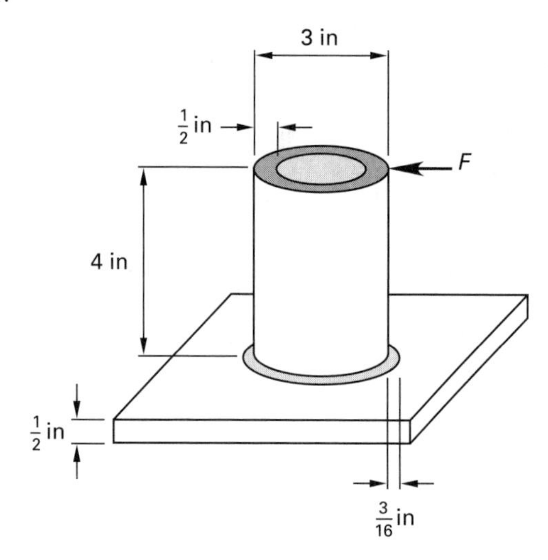

(A) 2900 lbf
(B) 3400 lbf
(C) 5700 lbf
(D) 17,000 lbf

22. An older, ideal, vapor refrigeration cycle using Freon-12 is used in an air conditioning application. The saturation temperature of the condenser is 90°F, and the evaporator temperature is 40°F. The horsepower needed to generate 3 tons of cooling is most nearly

(A) 0.57 hp
(B) 1.5 hp
(C) 2.8 hp
(D) 4.3 hp

23. A 20 in diameter suction line carries water at 7000 gpm. The water enters a pump at 100 psi and is pumped up a 30 ft high incline through a 12 in diameter discharge line. If the pressure at the top of the incline is 500 psi and the head loss through the 30 ft section is 10 ft, the horsepower that is supplied by the pump is most nearly

(A) 140 hp
(B) 750 hp
(C) 1200 hp
(D) 1700 hp

24. A 150 lbm cartop carrier measures 51 in long by 35 in wide by 18 in high. If a 3000 lbm car is driven at 65 mph into a 10 mph headwind, the added net motor power required when the carrier is used is most nearly

(A) 5.9 hp
(B) 13 hp
(C) 15 hp
(D) 36 hp

25. A jet of compressed air is used to bring a large gen-set (a turbine-powered emergency electrical generator) up to speed in 3 sec. The compressed air acts perpendicularly to the turbine's longitudinal axis of rotation. The effective torque moment arm (causing rotation) of the jet is 1.5 ft. The four-pole generator is direct-driven at 188.5 rad/sec. The compressed air is introduced at three times the tangential velocity of the turbine, but the air supply valve closes when the turbine has reached 110% of its steady-state rotational velocity. What is the initial Mach number of the compressed air used to start the gen-set?

(A) 0.41
(B) 0.54
(C) 0.62
(D) 0.75

26. Windows to be installed in a building are specified with glass having an R-value of 1.5 hr-ft^2-°F/Btu. Outdoor temperatures as low as −40°F are experienced in the winter, and the indoor space is to be heated to 70°F. Assuming indoor and outdoor window film coefficients of 1.46 Btu/hr-ft^2-°F and 6.0 Btu/hr-ft^2-°F respectively, the maximum relativity humidity that can be maintained without experiencing condensation on the inside of the glass is most nearly

(A) 31%
(B) 34%
(C) 37%
(D) 42%

27. A cylindrical tank is produced from layers of a low-ε glass, fiber-reinforced epoxy. The tank contains nitrogen gas at a pressure of 10 MPa. The tank has a mean diameter of 300 mm. The fiber orientation angles in various layers of the tank wall are at 55° with respect to the principal axes along the longitudinal and circumferential directions. There is no torsion on the tank. To limit the compressive stress to 115 MPa in the direction of fiber orientation, what is most nearly the minimum wall thickness?

(A) 8.7 mm
(B) 14 mm
(C) 21 mm
(D) 33 mm

28. A 500 g book lays at rest on a broken shelf that has a slope of 37° relative to the bookcase's level base. What is the minimum coefficient of static friction between the book and the shelf?

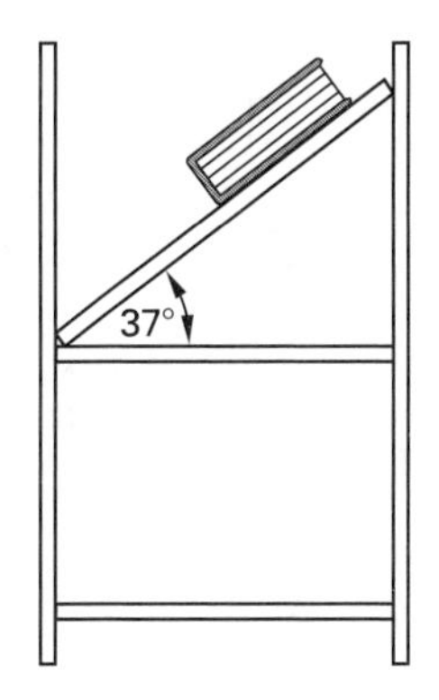

(A) 0.25
(B) 0.37
(C) 0.59
(D) 0.75

29. Air with a mass of 3 lbm is expanded polytropically behind a piston in a cylinder from an initial condition of 120 psia and 100°F to a final condition of 40 psia. Using a polytropic exponent of 1.6, the work done by the closed system per unit mass is most nearly

(A) -1.68×10^4 ft-lbf/lbm
(B) -5.60×10^3 ft-lbf/lbm
(C) 5.60×10^3 ft-lbf/lbm
(D) 1.68×10^4 ft-lbf/lbm

30. A heat exchanger pump receives water at 140°F and returns it at 95°F. The water is received and returned through nominal 3 in copper tubing (with an outside diameter of 3.5 in, inside diameter of 3.062 in, cross-sectional area of 7.370 in^2, and electrical resistivity of 10 Ω-cmil/ft). The pump is driven by a one-phase, 110 V AC, 5 hp motor that draws an average (RMS) current of 35 A. The engineer decides to place the copper tubing inside PVC pipe for electrical insulation, and use the copper tubing to carry the electrical current that will power the pump motor. The maximum distance that the pump and motor can be located from a 110 V source so that the voltage drop does not exceed 0.6 V is most nearly

(A) 870 ft
(B) 2500 ft
(C) 4900 ft
(D) 9000 ft

31. A 2 hp motor is used to stir a tank containing 60 lbm of water for 15 min. Assuming the process occurs at constant volume, the maximum possible rise in temperature is most nearly

(A) 1.4°F
(B) 21°F
(C) 58°F
(D) 130°F

32. What is the maximum value of A such that the shaft nose will never protrude from the left side of the bearing?

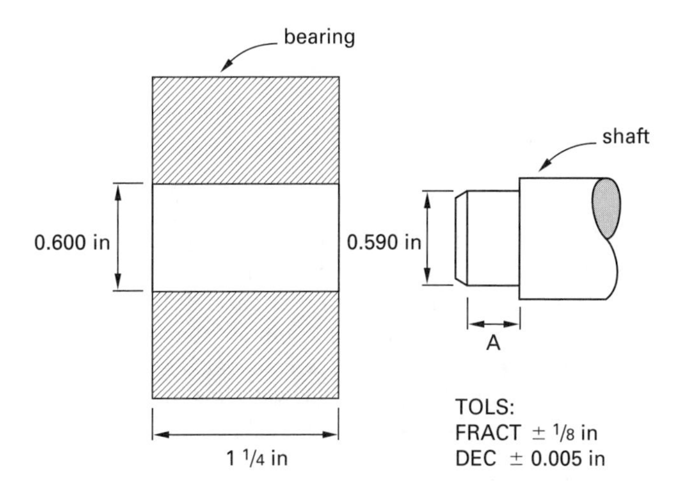

(A) 0.99 in
(B) 1 in
(C) 1.075 in
(D) 1.24 in

33. The rectangular wall of a furnace is made from 3 in fire-clay brick surrounded by 0.25 in of steel on the outside. 0.25 in diameter mild steel bolts connect the steel to the brick. The furnace is surrounded by 70°F air with a convection coefficient of 1.65 Btu/hr-ft^2-°F, while the inner surface of the brick is held constant at 1000°F. The outside surface temperature of the steel is most nearly

(A) 160°F
(B) 470°F
(C) 610°F
(D) 760°F

34. A parallel-flow tube-and-shell heat exchanger is designed using 1 in OD tubing. 40,000 lbm/hr of water at 55°F are used to cool 45,000 lbm/hr of a 95% ethyl alcohol solution (c_p = 0.9 Btu/lbm-°F) from 160°F to 110°F. If the overall coefficient of heat transfer based on the outer tube area is 75 Btu/hr-ft^2-°F, then the heat-transfer surface area of the heat exchanger is most nearly

(A) 520 ft^2
(B) 850 ft^2
(C) 920 ft^2
(D) 1100 ft^2

35. A steam turbine operates as a component of a Rankine cycle. Steam is supplied to the turbine at 1000 psia and 800°F. The turbine exhausts at 4 psia. The expansion is not reversible, and the exhaust is vapor at 100% quality. The thermal efficiency of the turbine is most nearly

(A) 59%
(B) 76%
(C) 85%
(D) 100%

36. A hydraulic press is used to compress soap cakes into shaped bars. The hydraulic fluid has an operating pressure of 900 psig and a flow rate of 3 gpm, and the press has an 85% efficient cylinder that travels at 0.25 ft/sec. The work done in compressing a tray of soap bars is 50 ft-lbf. Approximately how much is the thickness of each soap cake reduced in compression?

(A) 0.12 in
(B) 0.14 in
(C) 0.17 in
(D) 0.20 in

37. The gravimetric air/fuel ratio of methane burned with 125% theoretical air is most nearly

(A) 13:1
(B) 17:1
(C) 22:1
(D) 35:1

38. What type of fit is indicated by the manufacturing illustration shown?

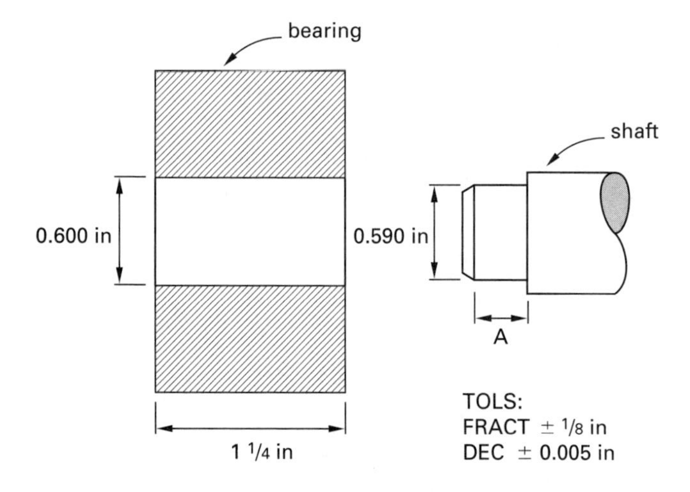

(A) clearance fit
(B) interference fit
(C) press fit
(D) tolerance fit

39. Water at 20°C flows through the pipe system (19 mm ID) shown at a rate of 185 L/min (0.185 m^3/min). If the loss coefficients for the 90° elbows and 45° elbow are 1.5 and 0.4, respectively, what is most nearly the sum of the minor (head) losses for the system?

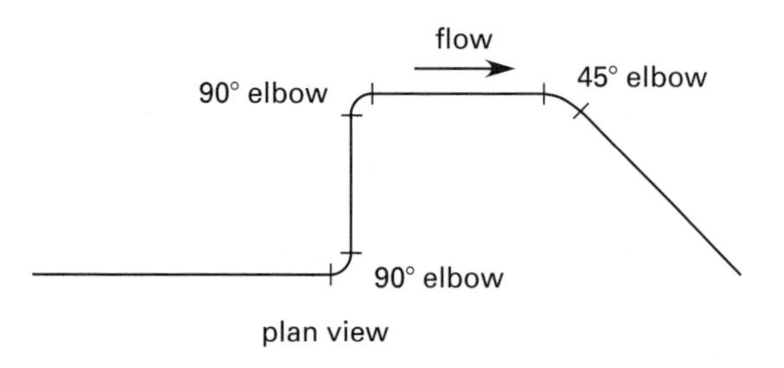

(A) 2.4 m
(B) 18 m
(C) 21 m
(D) 23 m

40. A solid copper sphere 200 mm in diameter is at a temperature of 450°C. It is dropped into a large tank of 75°C oil. If the average convective heat transfer coefficient is 880 $W/m^2 \cdot K$ and the oil is stirred uniformly at all times, most nearly how long after submersion does the sphere reach a temperature of 200°C?

(A) 5.3 s
(B) 160 s
(C) 910 s
(D) 4700 s

STOP!

DO NOT CONTINUE!

This concludes the Morning Session of the examination. If you finish early, check your work and make sure that you have followed all instructions. After checking your answers, you may turn in your examination booklet and answer sheet and leave the examination room. Once you leave, you will not be permitted to return to work or change your answers.

Afternoon Session
Instructions

In accordance with the rules established by your state, you may use textbooks, handbooks, bound reference materials, and any approved battery- or solar-powered, silent calculator to work this examination. However, no blank papers, writing tablets, unbound scratch paper, or loose notes are permitted. Sufficient room for scratch work is provided in the Examination Booklet.

You are not permitted to share or exchange materials with other examinees. However, the books and other resources used in this afternoon session do not have to be the same as were used in the morning session.

This portion of the examination is divided into three depth modules. You may select any one of the modules, regardless of your work experience. However, you may not "jump around" and solve questions from more than one module.

You will have four hours in which to work this session of the examination. Your score will be determined by the number of questions you answer correctly. There is a total of 40 questions in each depth module. All 40 questions in the module selected must be worked correctly in order to receive full credit on the exam. There are no optional questions. Each question is worth one point. The maximum possible score for this section of the examination is 40 points.

Partial credit is not available. No credit will be given for methodology, assumptions, or work written in your Examination Booklet.

Record all of your answers on the Answer Sheet. No credit will be given for answers marked in the Examination Booklet. Mark your answers with the pencil provided to you. Marks must be dark and must completely fill the bubbles. Record only one answer per question. If you mark more than one answer, you will not receive credit for the question. If you change an answer, be sure the old bubble is erased completely; incomplete erasures may be misinterpreted as answers.

If you finish early, check your work and make sure that you have followed all instructions. After checking your answers, you may turn in your Examination Booklet and Answer Sheet and leave the examination room. Once you leave, you will not be permitted to return to work or change your answers.

When permission has been given by your proctor, break the seal on the Examination Booklet. Check that all pages are present and legible. If any part of your Examination Booklet is missing, your proctor will issue you a new Booklet.

Do not work any questions from the Morning Session during the second four hours of this exam.

WAIT FOR PERMISSION TO BEGIN

Name: ______________________________
Last First Middle Initial

Examinee number: ______________

Examination Booklet number: ______________

Specialty Option to be graded: ______________

Principles and Practice of Engineering Examination

Afternoon Session Sample Examination

Depth Modules

Afternoon Session—HVAC and Refrigeration

41. (A) (B) (C) (D)	51. (A) (B) (C) (D)	61. (A) (B) (C) (D)	71. (A) (B) (C) (D)
42. (A) (B) (C) (D)	52. (A) (B) (C) (D)	62. (A) (B) (C) (D)	72. (A) (B) (C) (D)
43. (A) (B) (C) (D)	53. (A) (B) (C) (D)	63. (A) (B) (C) (D)	73. (A) (B) (C) (D)
44. (A) (B) (C) (D)	54. (A) (B) (C) (D)	64. (A) (B) (C) (D)	74. (A) (B) (C) (D)
45. (A) (B) (C) (D)	55. (A) (B) (C) (D)	65. (A) (B) (C) (D)	75. (A) (B) (C) (D)
46. (A) (B) (C) (D)	56. (A) (B) (C) (D)	66. (A) (B) (C) (D)	76. (A) (B) (C) (D)
47. (A) (B) (C) (D)	57. (A) (B) (C) (D)	67. (A) (B) (C) (D)	77. (A) (B) (C) (D)
48. (A) (B) (C) (D)	58. (A) (B) (C) (D)	68. (A) (B) (C) (D)	78. (A) (B) (C) (D)
49. (A) (B) (C) (D)	59. (A) (B) (C) (D)	69. (A) (B) (C) (D)	79. (A) (B) (C) (D)
50. (A) (B) (C) (D)	60. (A) (B) (C) (D)	70. (A) (B) (C) (D)	80. (A) (B) (C) (D)

Afternoon Session—Machine Design

81. (A) (B) (C) (D)	91. (A) (B) (C) (D)	101. (A) (B) (C) (D)	111. (A) (B) (C) (D)
82. (A) (B) (C) (D)	92. (A) (B) (C) (D)	102. (A) (B) (C) (D)	112. (A) (B) (C) (D)
83. (A) (B) (C) (D)	93. (A) (B) (C) (D)	103. (A) (B) (C) (D)	113. (A) (B) (C) (D)
84. (A) (B) (C) (D)	94. (A) (B) (C) (D)	104. (A) (B) (C) (D)	114. (A) (B) (C) (D)
85. (A) (B) (C) (D)	95. (A) (B) (C) (D)	105. (A) (B) (C) (D)	115. (A) (B) (C) (D)
86. (A) (B) (C) (D)	96. (A) (B) (C) (D)	106. (A) (B) (C) (D)	116. (A) (B) (C) (D)
87. (A) (B) (C) (D)	97. (A) (B) (C) (D)	107. (A) (B) (C) (D)	117. (A) (B) (C) (D)
88. (A) (B) (C) (D)	98. (A) (B) (C) (D)	108. (A) (B) (C) (D)	118. (A) (B) (C) (D)
89. (A) (B) (C) (D)	99. (A) (B) (C) (D)	109. (A) (B) (C) (D)	119. (A) (B) (C) (D)
90. (A) (B) (C) (D)	100. (A) (B) (C) (D)	110. (A) (B) (C) (D)	120. (A) (B) (C) (D)

Afternoon Session—Thermal and Fluids Systems

121. (A) (B) (C) (D)	131. (A) (B) (C) (D)	141. (A) (B) (C) (D)	151. (A) (B) (C) (D)
122. (A) (B) (C) (D)	132. (A) (B) (C) (D)	142. (A) (B) (C) (D)	152. (A) (B) (C) (D)
123. (A) (B) (C) (D)	133. (A) (B) (C) (D)	143. (A) (B) (C) (D)	153. (A) (B) (C) (D)
124. (A) (B) (C) (D)	134. (A) (B) (C) (D)	144. (A) (B) (C) (D)	154. (A) (B) (C) (D)
125. (A) (B) (C) (D)	135. (A) (B) (C) (D)	145. (A) (B) (C) (D)	155. (A) (B) (C) (D)
126. (A) (B) (C) (D)	136. (A) (B) (C) (D)	146. (A) (B) (C) (D)	156. (A) (B) (C) (D)
127. (A) (B) (C) (D)	137. (A) (B) (C) (D)	147. (A) (B) (C) (D)	157. (A) (B) (C) (D)
128. (A) (B) (C) (D)	138. (A) (B) (C) (D)	148. (A) (B) (C) (D)	158. (A) (B) (C) (D)
129. (A) (B) (C) (D)	139. (A) (B) (C) (D)	149. (A) (B) (C) (D)	159. (A) (B) (C) (D)
130. (A) (B) (C) (D)	140. (A) (B) (C) (D)	150. (A) (B) (C) (D)	160. (A) (B) (C) (D)

Afternoon Session
HVAC and Refrigeration

41. A refrigeration system using R-134a refrigerant operates at 33 psia on the low-pressure side and 100 psia on the high-pressure side. The 255 ton system delivers refrigerant vapor with 20°F superheat to the compressor. If the refrigerant on the high-pressure side is cooled to saturated liquid before expansion, the required refrigerant mass flow rate is most nearly

(A) 43,000 lbm/hr
(B) 48,000 lbm/hr
(C) 51,000 lbm/hr
(D) 55,000 lbm/hr

42. 7000 ft^3/min of recirculated air from a room conditioned to 75°F and 50% relative humidity are mixed with 2300 ft^3/min of outdoor air at 90°F db and 75°F wb prior to passing through a cooling coil and fan for distribution back to the room. The humidity ratio of the air entering the coil is most nearly

(A) 0.0093 lbm moisture/lbm dry air
(B) 0.011 lbm moisture/lbm dry air
(C) 0.014 lbm moisture/lbm dry air
(D) 0.015 lbm moisture/lbm dry air

43. A centrifugal pump with a net positive suction head requirement (NPSHR) of 12 ft water is required to deliver 80°F water from a lake to a storage reservoir. The lake and the reservoir are located near sea level. The reservoir surface is 17 ft above the lake surface. The pump itself is located near the surface of the storage reservoir. Friction and fitting losses are estimated to be 3.0 ft water. The net positive suction head available (NPSHA) for the pump is most nearly

(A) 13 ft water
(B) 14 ft water
(C) 16 ft water
(D) 20 ft water

44. A centrifugal pump is driven at 1300 rpm by a 10 hp motor and delivers 250 gpm of 85°F water against 75 ft water head. Assuming that the initial pump efficiency of 65% does not vary appreciably, the maximum flow the pump can deliver is most nearly

(A) 280 gpm
(B) 470 gpm
(C) 520 gpm
(D) 650 gpm

45. An air conditioning system is being designed for a 40,000 ft^2 casino which will operate 24 hr/day. The space is completely inside a larger building and has no windows or walls exposed to direct sun on the exterior of the casino. The designer anticipates that the lighting system will require approximately 3.75 W/ft^2. The gaming equipment, cameras, and miscellaneous equipment require an additional 80 kW. The steam tables for a large all-day buffet contribute an additional 50,000 Btu/hr of latent load. The maximum density for patrons is estimated to be 120 people/1000 ft^2. It is desired to maintain space conditions at 73°F and 45% relative humidity. The total space cooling load due to internal heat gain is most nearly

(A) 890,000 Btu/hr
(B) 1,900,000 Btu/hr
(C) 2,400,000 Btu/hr
(D) 3,000,000 Btu/hr

46. An office space measures 12 ft × 12 ft and has a 9 ft ceiling. The cooling load has been determined to be 3000 Btu/hr. A single ceiling diffuser will be installed in the center of the room ceiling. To achieve good occupant comfort and a maximum air distribution performance index (ADPI), a diffuser should be selected with a throw of most nearly

(A) 5.0 ft
(B) 6.0 ft
(C) 10 ft
(D) 12 ft

47. After a driver parked and exited her automobile, the air-conditioned interior was 70°F at 40% relative humidity. After the car sat in the driveway overnight, the interior temperature dropped to 50°F, bringing the interior's relative humidity to most nearly

(A) 40%
(B) 60%
(C) 80%
(D) 100%

48. Refrigerant at −21°C passes through a thin cylindrical tube 15 mm in diameter. Ambient air passing over the pipe is cooled such that the heat extraction rate per meter pipe length from the ambient air is 73.0 W. Under steady-state conditions, the heat convection coefficient is 80 $W/m^2 \cdot K$. What is most nearly the temperature of the ambient air? Neglect conductive and radiative heat transfer.

(A) −250°C
(B) −40°C
(C) −2.0°C
(D) 2.0°C

49. A fan supplies 4500 ft^3/min of air through a 240 ft long rectangular duct. The duct dimensions are 18 in × 24 in. If the duct has a friction factor of 0.016 and a roughness of 0.0003 ft, the total static pressure drop due to friction is most nearly

(A) 0.15 in water
(B) 0.31 in water
(C) 0.45 in water
(D) 0.60 in water

50. The water pipes in a vented crawl space are uninsulated. The space shares 320 ft^2 of exterior wall with the outdoors and 645 ft^2 of floor with a space above that is heated to 72°F. The overall heat transfer coefficients (U-factors) for the crawl space wall and floor above are

$$U_{\text{floor}} = 0.05 \text{ Btu/hr-ft}^2\text{-°F}$$

$$U_{\text{wall}} = 0.15 \text{ Btu/hr-ft}^2\text{-°F}$$

Outdoor air at 900 ft^3/hr infiltrates the crawl space through the vents. The pipes will be safe from freezing down to an outdoor temperature of most nearly

(A) −12°F
(B) 2.0°F
(C) 12°F
(D) 32°F

51. Water enters an evaporative cooling tower as a saturated liquid at 180°F. 8% of the water evaporates as saturated vapor. After this process, the temperature of the remaining water is most nearly

(A) 81°F
(B) 86°F
(C) 94°F
(D) 101°F

52. The water surface in a well is 90 ft below a house. A storage tank with its water surface 50 ft above the house will provide gravity water flow to the house. A submersible pump is capable of delivering 6 gpm from the well to the tank. The system is powered directly by a solar-electric array that has an overall efficiency of 12% and regularly intercepts solar radiation of a magnitude of at least 28 W/ft^2 on a cloudless day. The pump efficiency is 60%, and piping has been oversized to such an extent that friction and fitting losses are negligible. The minimum required area for the solar array operating on a cloudless day is most nearly

(A) 6.0 ft^2
(B) 16 ft^2
(C) 50 ft^2
(D) 79 ft^2

53. To reduce the load on a chiller plant, an air washer recirculating 57°F water is used to evaporatively precool 20,000 ft^3/min outdoor air. The outdoor air is introduced at 92°F db and 57°F wb. If the saturation efficiency of the process is 84%, the cooling requirement reduction is most nearly

(A) 45 tons
(B) 49 tons
(C) 53 tons
(D) 63 tons

54. An electric unit heater is needed to heat a room on the second floor of a three-story building. The room has no windows. The 800 ft^2 external wall is made up of 2 in of polystyrene rigid insulation sandwiched between 8 in brick and 5/8 in gypsum board. The outdoor winter design temperature is 10°F, and the room is to be maintained at 74°F. Using a 25% safety factor, the required heating capacity for the unit heater is most nearly

(A) 1.5 kW
(B) 1.9 kW
(C) 2.4 kW
(D) 2.8 kW

55. 80°F air is introduced to a 3 ft × 4 ft (face-area dimensions) cooling coil at a velocity of 450 ft/min and is cooled to 56°F. The coil cooling process has a grand sensible heat ratio (GSHR) of 0.70. The total coil load is most nearly

(A) 120,000 Btu/hr
(B) 140,000 Btu/hr
(C) 180,000 Btu/hr
(D) 200,000 Btu/hr

56. In a certain system, 2530 gpm of 83°F condenser water are supplied to an 840 ton chiller for heat rejection and are returned to the cooling tower at 92°F. The coefficient of performance (COP) of the chiller is most nearly

(A) 3.8
(B) 4.5
(C) 6.0
(D) 7.7

57. Which of the following centrifugal fan impeller designs typically exhibits the highest efficiency?

(A) backward curved
(B) forward curved
(C) backward inclined
(D) airfoil

58. A 6 ft^3 gas bottle holds 5 lbm of compressed nitrogen gas at 100°F. Gas is released until the bottle pressure reaches 150 lbf/in^2. What is most nearly the amount of gas released?

(A) 0.030 lbm
(B) 0.80 lbm
(C) 1.7 lbm
(D) 4.2 lbm

59. A store in a shopping mall is to be maintained at 75°F and 45% relative humidity with supply air of 55°F and 30% relative humidity. The space cooling load is 73,000 Btu/hr sensible and 26,000 Btu/hr latent at outdoor design conditions of 94°F db and 72°F wb. The ventilation requirement is 850 ft^3/min. The coil load due to the ventilation air is most nearly

(A) 33,000 Btu/hr
(B) 56,000 Btu/hr
(C) 75,000 Btu/hr
(D) 137,000 Btu/hr

60. The reheat coil in a variable air volume (VAV) terminal box is being replaced. The maximum airflow capacity of the box is 2400 ft^3/min. A minimum stop setting of 30% (of the maximum flow) has been established to maintain the required ventilation when cooling loads are at a minimum. The supply air temperature for the building system is a function of the outside air temperature, according to the reset graph shown. During the winter, the outdoor design temperature of 10°F and the indoor space temperature of 72°F result in a space heat loss of 45,000 Btu/hr. The minimum capacity of the reheat coil is most nearly

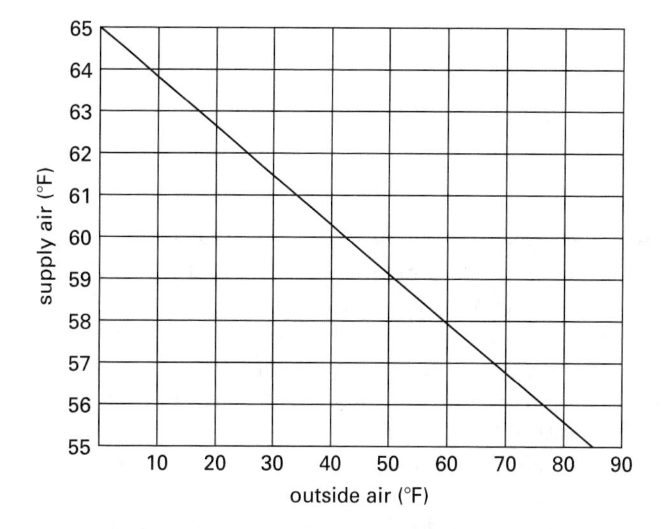

(A) 45,000 Btu/hr
(B) 47,000 Btu/hr
(C) 51,000 Btu/hr
(D) 66,000 Btu/hr

61. In a particular process, valves at each end of an 80 ft section of 2 in schedule-40 pipe are closed instantaneously, and the saturated steam (at atmospheric pressure) in the pipe begins to cool. The average heat transfer over the cooling period is 6200 Btu/hr. There is no heat loss through the valves at the ends of the pipe. The time it takes for the steam to cool to the ambient temperature of 60°F is most nearly

(A) 0.01 min
(B) 1 min
(C) 4 min
(D) 10 min

62. The ventilation system for a new 5000 ft^2 theater auditorium should provide treated outdoor air of acceptable quality at a rate of most nearly

(A) 2300 ft^3/min
(B) 6400 ft^3/min
(C) 11,000 ft^3/min
(D) 16,000 ft^3/min

63. A simple Rankine cycle operates between superheated steam entering a turbine at 1200°F and 700 psia, and entering a pump at 2 psia. The cycle's maximum possible efficiency is

(A) 27%
(B) 31%
(C) 39%
(D) 43%

64. A single-stage chiller circulates 117,000 lbm/hr of R-22 refrigerant and operates with a 90°F condensing temperature and 10°F evaporating temperature. Saturated refrigerant vapor enters the compressor with no superheat, and saturated liquid refrigerant leaves the condenser with no subcooling. Heat is rejected to condenser water that enters the condenser at 85°F and leaves at 95°F. The rated coefficient of performance (COP) is 5.5 under these conditions. The condenser water flow required for heat rejection is most nearly

(A) 560 gpm
(B) 1600 gpm
(C) 1900 gpm
(D) 2200 gpm

65. An expansion tank is being provided for a steel-pipe chilled water distribution system. The system has a volume of 1500 gallons and must be able to operate between temperature extremes of 50°F and 105°F. The minimum and maximum tank pressures are 10 psig and 23 psig, respectively. The steel pipe has a thermal linear expansion coefficient of 6.5×10^{-6} in/in-°F. If the tank is to be a closed type, with air/water interface, the minimum tank size is most nearly

(A) 20 gal
(B) 30 gal
(C) 50 gal
(D) 70 gal

66. Recirculated air at 7000 ft^3/min, pulled from a room conditioned to 75°F and 50% relative humidity, is mixed with 2300 ft^3/min of outdoor air at 90°F db and 75°F wb prior to passing through a cooling coil and a centrifugal fan for distribution back to the room. The coil has a total cooling capacity of 300,000 Btu/hr and a sensible cooling capacity of 235,000 Btu/hr. The fan power at the shaft is 4.2 BHP. The temperature of the air leaving the coil is most nearly

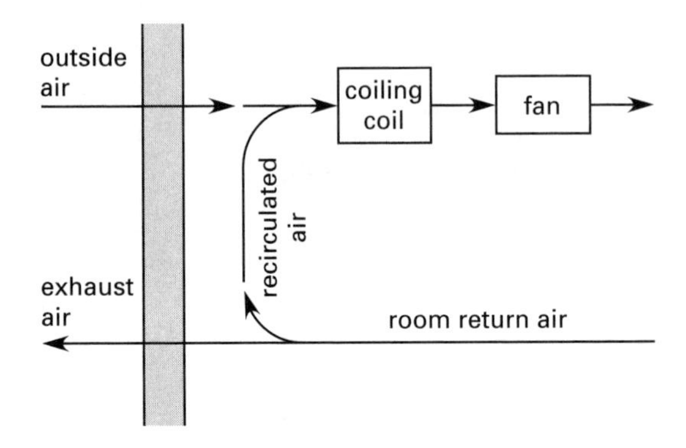

(A) 49°F
(B) 55°F
(C) 64°F
(D) 90°F

67. In the psychrometric process diagram shown, the process identified as 3-4 is best described as

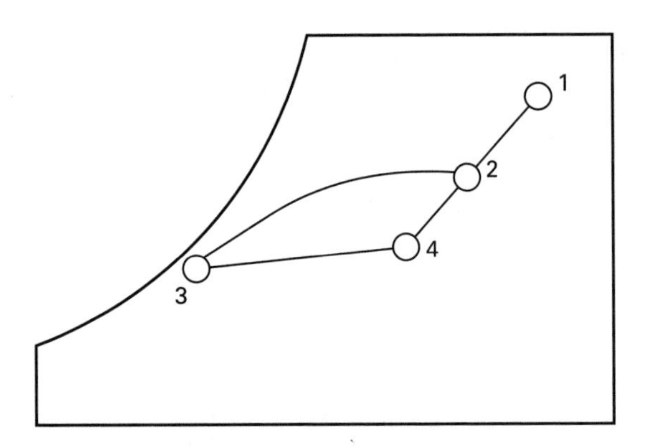

(A) evaporative cooling
(B) sensible and latent cooling
(C) dehumidification
(D) sensible and latent heating

68. The chiller system in a 1600 ft^2 (floor area) refrigeration machinery room contains 150 lbm of refrigerant. The room has 1.5 ft^2 of equivalent leakage area to the outside and an estimated heat gain of 15,000 Btu/hr. The design outside air temperature is 90°F. According to the Uniform Mechanical Code, refrigeration machinery rooms must be provided with dedicated mechanical exhaust with the capacity to achieve all of the following.

I. continuous maintenance of the room at 0.05 in water negative relative to adjacent spaces, Δp, with equivalent leakage area, A_e, in ft^2, as calculated by

$$\dot{Q}_1 = 2610 A_e \sqrt{\Delta p}$$

II. continuous airflow of 0.5 ft^3/min per gross ft^2, A_{gf}, within the room as calculated by

$$\dot{Q}_2 = 0.5 A_{gf}$$

III. temperature rise, ΔT, within the refrigeration machinery room due to heat dissipation, $\dot{q}$, in Btu/hr to a maximum of 104°F, limited by

$$\dot{Q}_3 = \frac{\sum \dot{q}}{1.08 \Delta T}$$

IV. emergency purge of escaping refrigerant mass G, in lbm, as calculated by

$$\dot{Q}_4 = 100\sqrt{G}$$

The dedicated mechanical exhaust required for the room is most nearly

(A) 800 ft^3/min
(B) 900 ft^3/min
(C) 1000 ft^3/min
(D) 1200 ft^3/min

69. A water-cooled condenser heats cooling water from 65°F to 95°F. 2200 lbm/hr of saturated water vapor enters the condenser at 4 psia and exits as saturated liquid. The mass flux of the exiting cooling water is most nearly

(A) 34,000 lbm/hr
(B) 66,000 lbm/hr
(C) 74,000 lbm/hr
(D) 83,000 lbm/hr

70. Published throw information for supply air grilles and ceiling diffusers is based on the tendency of air to adhere to the ceiling under the right conditions. This tendency is called the

(A) Coanda effect
(B) Bernoulli effect
(C) Darcy effect
(D) Colebrook effect

71. A sun room, which has an exposed 6 in concrete floor slab measuring 20 ft × 30 ft, has been heated passively by solar radiation to an average temperature of 82°F by nightfall. The room thermostat setting is kept at 60°F. Assuming a floor slab convective heat transfer coefficient of 1.4 Btu/hr-ft^2-°F, the average temperature change of the slab after 1 hr of cooling is most nearly

(A) 1°F
(B) 2°F
(C) 4°F
(D) 5°F

72. A 22-unit apartment complex is to be served by a central hot water system. The units will each contain two lavatories, one bathtub, one shower, one kitchen sink, and one dishwasher. The American Society of Heating, Refrigerating, and Air Conditioning Engineers (ASHRAE) publishes the hot-water demand (gallons per hour per fixture) for these fixtures as

fixture	demand (gal/hr)
lavatory	2
bathtub	20
shower	30
kitchen sink	10
dishwasher	15

ASHRAE also recommends the use of a 0.30 demand factor and a 1.25 storage factor. The appropriate size hot-water storage tank, based upon meeting a probable 1 hr demand, is most nearly

(A) 520 gal
(B) 650 gal
(C) 1700 gal
(D) 2200 gal

73. An east-facing vertical window at a latitude of 40 degrees north has an area of 12 ft^2. The shading coefficient for the window is 0.87. The overall heat transfer coefficient is 1.2 Btu/hr-ft^2-°F. The table shown gives solar heat gain factors for 40 degrees north.

	solar time	solar heat gain factor, Btu/hr-ft^2								solar time
	a.m.	N	NE	E	SE	S	SW	W	NW	p.m.
May	0500	0	2	2	0	0	0	0	0	1900
	0600	35	127	140	70	11	11	11	11	1800
	0700	27	164	208	130	21	20	20	20	1700
	0800	26	148	218	163	30	26	26	26	1600
	0900	30	104	198	176	52	29	31	31	1500
	1000	35	53	149	164	82	34	35	35	1400
	1100	35	39	80	131	104	42	35	35	1300
	1200	38	36	41	83	112	81	39	39	1200
half day total		214	665	1023	880	357	199	175	174	
June	0500	9	20	21	5	2	2	2	2	1900
	0600	47	144	150	71	12	12	12	12	1800
	0700	36	171	206	121	21	20	20	22	1700
	0800	31	155	215	153	30	28	28	28	1600
	0900	32	113	191	160	44	33	31	31	1500
	1000	34	64	146	149	70	35	35	36	1400
	1100	37	41	80	115	89	40	39	37	1300
	1200	37	37	42	73	96	73	40	37	1200
half day total		252	735	1037	819	314	203	187	186	

On a day in May at solar time 0800, the indoor temperature is 75°F and the outdoor temperature is 42°F. The total instantaneous heat gain for this window is most nearly

(A) 480 Btu/hr
(B) 1800 Btu/hr
(C) 2300 Btu/hr
(D) 2900 Btu/hr

74. A building's facility manager has proposed that an air-to-air heat exchanger be installed between a system's outside air and exhaust air streams to provide "free" outside air preheat. On a winter design day, the 10°F db outdoor air requirement is 12,500 ft^3/min, and 10,000 ft^3/min of 72°F building air is exhausted. The manufacturer's published heat exchanger effectiveness is 0.64. Under these conditions, the rate of energy savings is most nearly

(A) 430,000 Btu/hr
(B) 490,000 Btu/hr
(C) 550,000 Btu/hr
(D) 850,000 Btu/hr

75. A property owner has installed an air-cooled chiller on the side of his building, as shown. The neighboring property owner has complained that the unit is too noisy, particularly at night. The manufacturer's data states that the unit has a sound power rating of 80.0 dB. The local municipal code requires that the sound pressure level be no greater than 52 dB at the property line. The sound pressure level at the property line is most likely

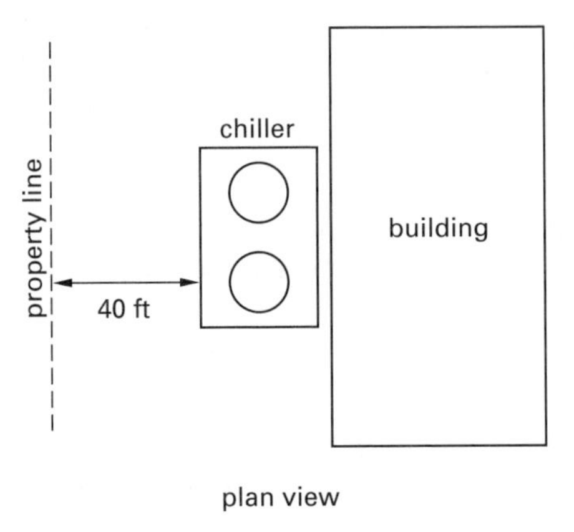

(A) 48 dB
(B) 51 dB
(C) 53 dB
(D) 57 dB

76. The employees of an insurance company insist that their office is too cold. When the temperature was measured to be 72°F, they responded, "it feels colder than that in here." The single-pane window has been measured to have a surface temperature of 35°F, compared to the walls, floor, and ceiling which average 65°F. The calculated angle factor between a typical workstation and the window is 0.04. Assuming 50 ft/min air velocity, the operative temperature at the typical workstation is most nearly

(A) 62°F
(B) 64°F
(C) 68°F
(D) 70°F

77. The specific heat for air at constant pressure is 0.240 Btu/lbm-°R. The specific heat for air at constant volume is 0.171 Btu/lbm-°R. The work needed to compress 10 lbm of air isentropically from atmospheric pressure and a temperature of 55°F to a pressure of 720 psia is most nearly

(A) 590 Btu
(B) 820 Btu
(C) 1600 Btu
(D) 1800 Btu

78. During a routine inspection, a plant engineer discovers a section of bare overhead steam pipe. Upon checking the plant's maintenance records, the engineer learns that a leaking steam trap had recently been repaired, and the saturated insulation had been removed from the pipe but never replaced.

The properties of the pipe are as follows.

length of bare pipe section	120 ft
pipe material	1% carbon steel
pipe size	1½ in BWG 16 gage
pipe mounting	ceiling pipe hangers

Saturated steam at atmospheric pressure flows through the pipe at a rate high enough to prevent substantial condensation. The average inside heat transfer coefficient is 1500 Btu/hr-ft^2-°F. The average outside heat transfer coefficient of the bare pipe in still air is 2.0 Btu/hr-ft^2-°F. The air in the plant is at 60°F and 14.7 psia and is normally still. The pipe temperature is too low to consider the effects of radiation. The rate of heat loss from the bare pipe is most nearly

(A) 14,300 Btu/hr
(B) 17,500 Btu/hr
(C) 18,900 Btu/hr
(D) 20,800 Btu/hr

79. A fire sprinkler system uses 1 in nominal diameter schedule-40 black steel pipe. The sprinklers are located 10 ft apart. The minimum pressure at any sprinkler is 10 psig. All sprinklers have standard ½ in orifices with discharge coefficients of 0.75. In a particular event, the last three sprinklers on a branch line are fully open simultaneously. Disregarding velocity pressure, the second branch sprinkler from the end will discharge most nearly

(A) 11 gpm
(B) 15 gpm
(C) 19 gpm
(D) 23 gpm

80. A home owner plans to have a contractor install a standard furnace with a rated annual fuel utilization efficiency (AFUE) of 78%. It is estimated that the furnace will consume 1500 therms of natural gas per year. The contractor has offered to install a high-efficiency condensing furnace with an AFUE of 92% for an additional $800. Assuming a constant cost of natural gas of $0.85 per therm, the simple payback period of this additional investment will be most nearly

(A) 2 yr
(B) 4 yr
(C) 5 yr
(D) 12 yr

STOP!

DO NOT CONTINUE!

This concludes the Afternoon Session of the examination. If you finish early, check your work and make sure that you have followed all instructions. After checking your answers, you may turn in your examination booklet and answer sheet and leave the examination room. Once you leave, you will not be permitted to return to work or change your answers.

Afternoon Session
Machine Design

81. A rotary pendulum comprises a vertical axis, three steel balls spinning around that axis, and an ideal torsional spring resisting the motion. Arms supporting the balls are rigid and massless, the distance from the vertical axis to ball centers is 0.75 in, and the torsional spring constant is 2.857×10^{-3} in-lbf/rad. The steel ball diameter and density are 1.00 in and 0.284 lbm/in^3, respectively. What is most nearly the natural torsional frequency of the pendulum?

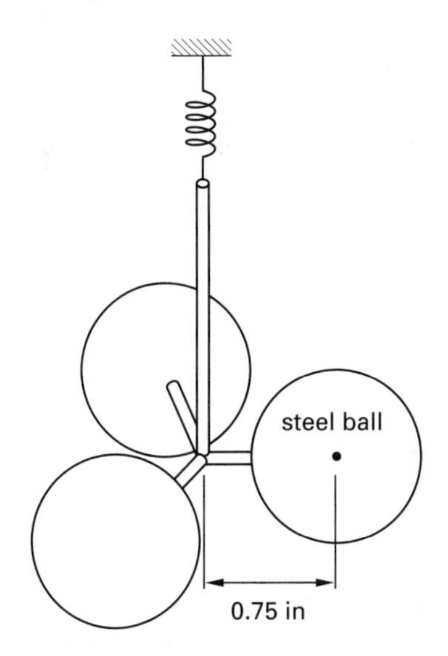

(A) 0.19 Hz
(B) 0.31 Hz
(C) 0.35 Hz
(D) 0.53 Hz

82. A 500 lbm artillery projectile is launched with initial velocity of 2000 ft/sec, launch angle of 25° from the horizontal, and impact elevation of 2500 ft below the launch elevation. Neglecting air resistance, what is most nearly the horizontal travel distance?

(A) 10 mi
(B) 18 mi
(C) 19 mi
(D) 43 mi

83. A 500 lbf load is eccentrically applied to a 3 in diameter circle of six bolts as shown. What is most nearly the maximum bolt shear load?

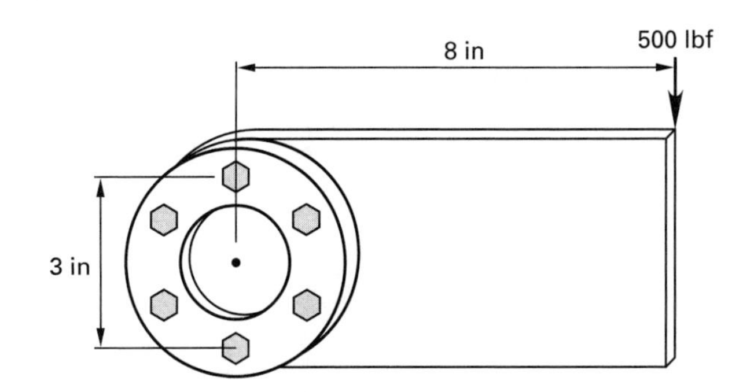

(A) 370 lbf
(B) 450 lbf
(C) 490 lbf
(D) 510 lbf

84. A titanium tube is loaded in torsion. The 2 ft long tube has a 2 in outside diameter and a 0.125 in wall thickness. The tensile modulus of elasticity is 15.9×10^6 lbf/in^2. The maximum applied torque is 50,000 in-lbf. The maximum angular deflection is most nearly

(A) 0.026 rad
(B) 0.12 rad
(C) 0.31 rad
(D) 0.56 rad

85. A radial ball bearing has a basic load rating of 5660 lbf based on life of 1,000,000 cycles. The bearing supports a 5000 lbf radial load and a 4000 lbf thrust load. The radial loading factor is 0.56, the thrust loading factor is 1.31, the outer ring rotates, and there is no shock loading. What is most nearly the expected service life?

(A) 280×10^3 cycles
(B) 350×10^3 cycles
(C) 710×10^3 cycles
(D) 840×10^3 cycles

86. A two-bar linkage is rotating about a fixed point O with a constant angular velocity of 30 rad/sec as shown. At a particular moment, bar OA forms an angle of 30° with the horizon, and the velocity of point B is 5 in/sec directed horizonally to the left.

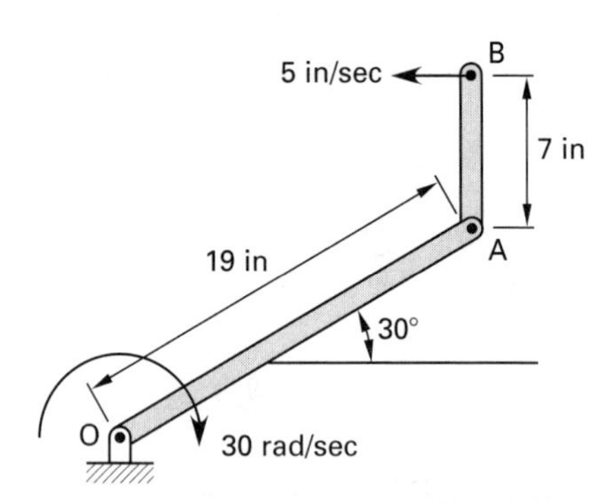

The instantaneous angular velocity of bar AB is most nearly

(A) 33 rad/sec
(B) 40 rad/sec
(C) 67 rad/sec
(D) 80 rad/sec

87. A spur gear has a pitch diameter of 4.00 in, face width of 1.67 in, 34 teeth, and a Lewis form factor of 0.138. Bending stress at the tooth root is limited to 40,000 lbf/in^2. Disregarding stress concentration, what is most nearly the allowable tangential load per tooth?

(A) 3400 lbf
(B) 10,700 lbf
(C) 24,900 lbf
(D) 33,600 lbf

88. A simply supported rectangular-section steel beam is 3.0 in wide, 0.32 in thick, and 35.0 in long. Bending stress is limited to 29,000 lbf/in^2. Deflection is limited to 0.50 in. Neglect buckling. The maximum point load that can be supported at the beam midpoint is most nearly

(A) 65 lbf
(B) 85 lbf
(C) 140 lbf
(D) 170 lbf

89. A helical compression spring made of 0.225 in diameter steel wire has a 2.20 in OD and a 6 in free length. Allowable shear stress is 95,000 lbf/in^2. The spring is dynamically compressed to 150 lbf. What is most nearly the factor of safety?

(A) 0.81
(B) 1.10
(C) 1.20
(D) 1.40

90. A cantilevered beam 40 m long is supported at its left end and 8 m from its right end, as shown. The beam bears a uniform load of 30 N/m and a concentrated load of 400 N applied 6 m from its right end. What is most nearly the vertical shear acting on the beam at point C?

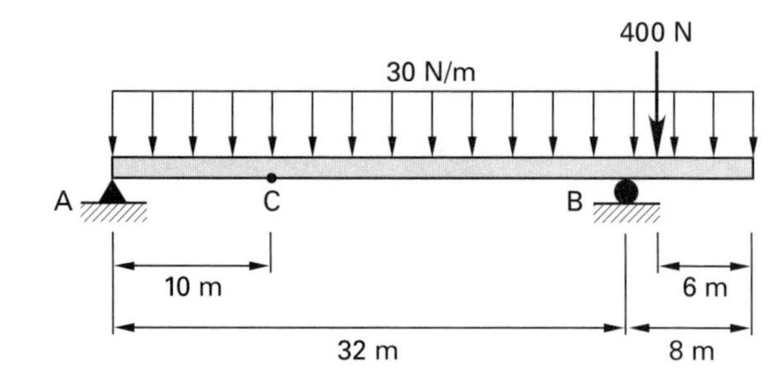

(A) 130 N
(B) 430 N
(C) 750 N
(D) 1200 N

91. A tubular steel shaft (with a modulus of elasticity of 30.5 × 10^6 psi) has a 2 in OD and a 0.125 in wall thickness. The shaft supports two vertically loaded power transmission components, A and B, as shown. What is most nearly the shaft's lowest critical speed?

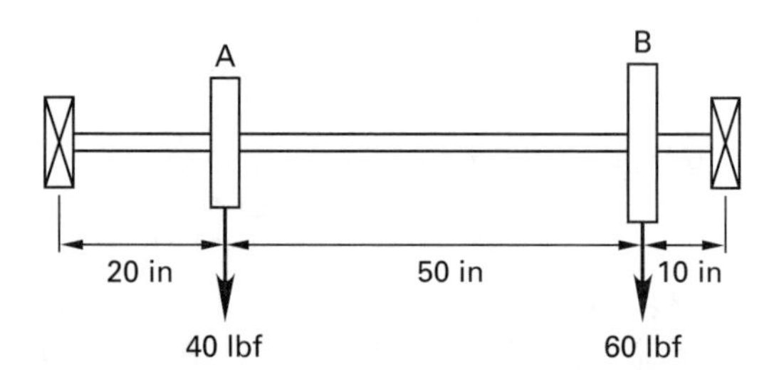

(A) 590 rpm
(B) 940 rpm
(C) 1100 rpm
(D) 1300 rpm

92. A 3.0 in diameter hollow steel shaft is subjected to a static 10,000 in-lbf bending moment and a static 15,000 in-lbf torsional moment. The material yield stress is 80,000 lbf/in^2, and the ultimate stress is 120,000 lbf/in^2. The design factor of safety is 2. The minimum internal diameter is most nearly

(A) 1.9 in
(B) 2.1 in
(C) 2.7 in
(D) 2.9 in

93. The bolted joint shown is designed to carry a shear force of 14,000 lbf and to fail by bolt shear before the force exceeds 20,000 lbf. Characteristics of candidate bolts are listed in the table. There are no threads in the shear plane. Which size bolt best meets the design specifications?

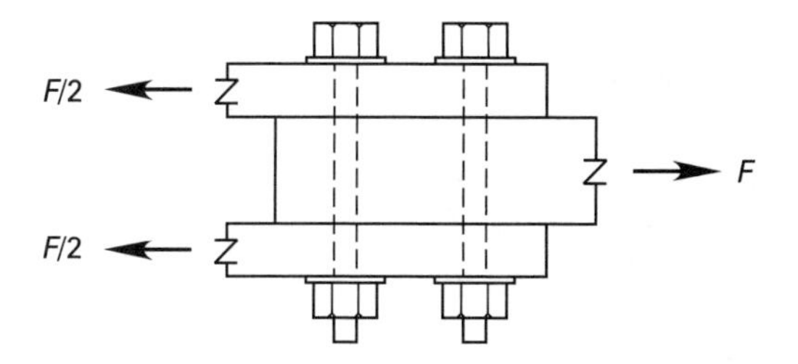

size (in)	major diameter (in)	minor diameter (in)	yield strength (lbf/in^2)	tensile strength (lbf/in^2)
1/4	0.2500	0.1887	57,500	74,500
3/8	0.3750	0.2983	57,500	74,500
1/2	0.5000	0.4056	57,500	74,500
5/8	0.6250	0.5135	57,500	74,500

(A) 1/4 in
(B) 3/8 in
(C) 1/2 in
(D) 5/8 in

94. A nut tightened on a 1″-8UNC-2A bolt develops the full 105,000 lbf/in^2 proof stress. The average radius of nut friction forces is 0.625 in, and all friction coefficients are 0.15. Bolt characteristics are as listed.

pitch diameter	0.9188 in
minor diameter	0.8512 in
lead angle	2.48°
thread half angle	30°
minor diameter area	0.551 in^2
tensile stress area	0.606 in^2

What amount of torque is required to tighten the nut to full proof stress?

(A) 1000 ft-lbf
(B) 1300 ft-lbf
(C) 1500 ft-lbf
(D) 2400 ft-lbf

95. The four-bar linkage in the illustration represents a mechanism. When the input angle ϕ is 120°, what is most nearly the output angle θ?

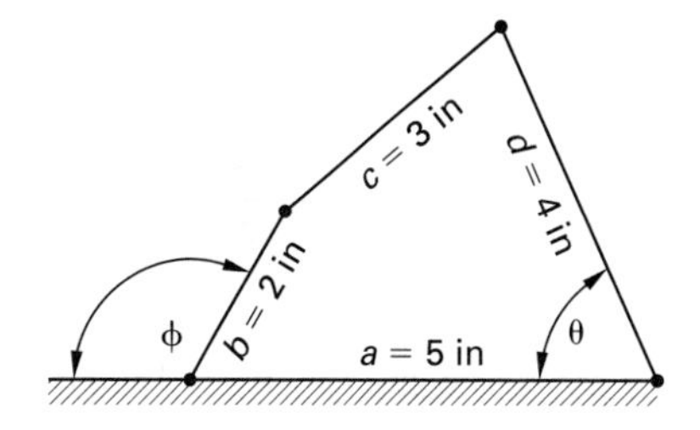

(A) 55°
(B) 65°
(C) 75°
(D) 85°

96. Design engineers must select a structural polymer from the following list of materials. One critical dimension, fabricated to 10.000 in at 25°C, must not exceed 10.045 in at the operating temperature of 80°C. Material weight must be less than half that of aluminum, which has a specific gravity of 2.71. Which is the best material for the application?

material	yield strength (lbf/in^2)	specific gravity	coefficient of thermal expansion (in/in-°C)	maximum operating temperature (°F)
I	4500	1.05	90×10^{-6}	200
II	10,000	1.40	70×10^{-6}	220
III	8000	1.10	80×10^{-6}	170
IV	9300	1.20	70×10^{-6}	260

(A) I
(B) II
(C) III
(D) IV

97. An element of a machine component has the static stress state shown. The material is anisotropic with tensile yield strength of 45,000 lbf/in^2. The shear yield strength is 10,000 lbf/in^2. What is most nearly the factor of safety?

$$\sigma_x = 10{,}000 \text{ lbf/in}^2$$
$$\sigma_y = 15{,}000 \text{ lbf/in}^2$$
$$\tau_{xy} = \tau_{yx} = 5000 \text{ lbf/in}^2$$

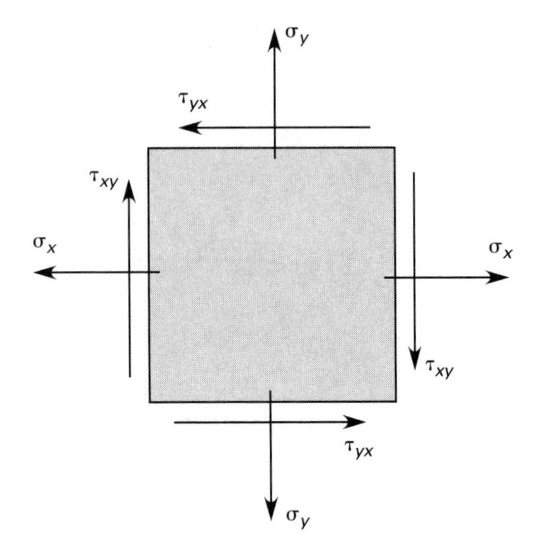

(A) 1.8
(B) 2.5
(C) 3.0
(D) 3.8

98. A machine may be upgraded or replaced. The annualized costs of each option are compared in the table.

	upgrade	replace
annual maintenance	\$500	\$100
initial cost	\$9000	\$40,000
future salvage value	\$10,000 @ year 20	\$15,000 @ year 25
present salvage value	\$13,000	–
interest rate	8%	8%

The uniform annual cost difference between these alternatives is most nearly

(A) \$360
(B) \$750
(C) \$1100
(D) \$2500

99. A flywheel has a 2 in thick rim and 36 in outside diameter as shown. The rim material is cast iron with a density of 0.26 lbm/in^3 and an ultimate strength of 100,000 lbf/in^2. The design safety factor is 10. What is most nearly the maximum rotational design speed?

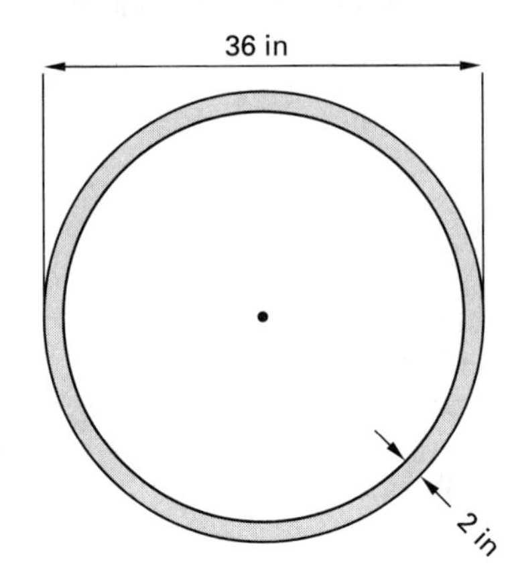

(A) 1900 rpm
(B) 2200 rpm
(C) 3800 rpm
(D) 5100 rpm

100. Extensive completely reversed load testing of a nonferrous machine element results in the endurance strength chart shown. A single 9 min load cycle is described in the table. What is most nearly the element fatigue life?

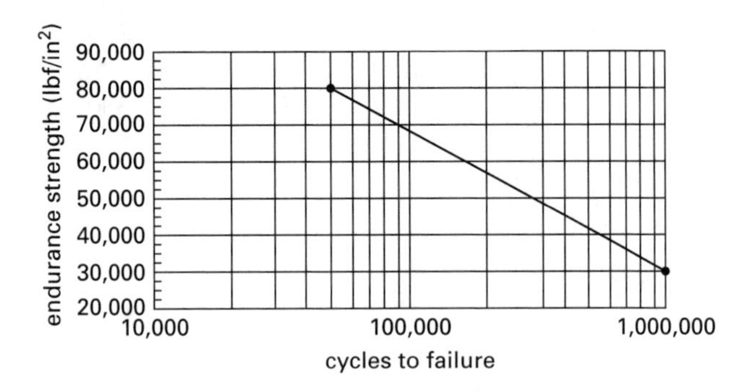

load (lbf/in^2)	duration (min)
±80,000	1
±50,000	3
±30,000	5

(A) 5500 hr
(B) 6200 hr
(C) 26,000 hr
(D) infinite

101. A 2 in long plastic push rod has a rectangular section measuring 1.0 in × 0.7 in. One end is rigidly fixed in place, while the other end is not supported in any direction. The material's tensile modulus is 420,000 lbf/in^2, and the 2% offset yield strength is 10,000 lbf/in^2. The critical buckling load is most nearly

(A) 5100 lbf
(B) 6100 lbf
(C) 6700 lbf
(D) 7400 lbf

102. A manufacturing process follows a standard normal distribution about the mean. Process results within $\pm 3\sigma$ of the mean are accepted; results beyond these limits are reworked or discarded. The probability that the process will produce an accepted result is most nearly

(A) 68.3%
(B) 95.4%
(C) 99.7%
(D) 99.9%

103. A steel pressure vessel has a 24 in outside diameter, 0.625 in thick shell and elliptical head, 0.5 in thick hemispherical head, and $^1/_{16}$ in corrosion allowance. Welds are spot radiographed, and operating temperature is 850°F. Allowable stresses versus temperature and weld joint efficiencies are given in the tables. What is most nearly the maximum design pressure?

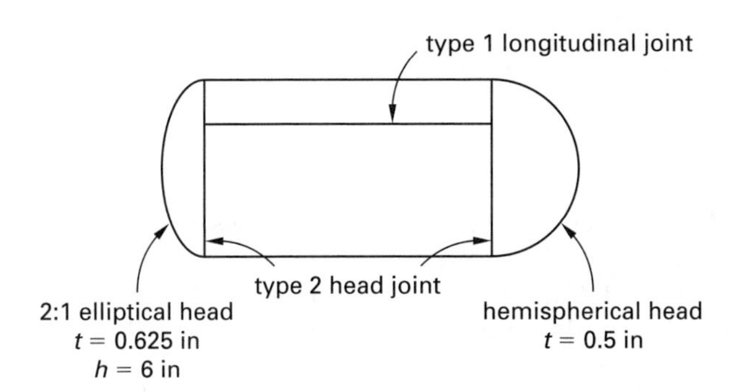

temperature (°F)	allowable stress (lbf/in^2)
700	16,700
750	13,900
800	11,400
850	8700
900	5900

structures	weld type	radiography full	radiography spot	radiography none
shells and hemispherical heads	1	1.00	0.85	0.70
shells and hemispherical heads	2	0.90	0.80	0.65
nonhemispherical heads	any	1.00	1.00	0.85

(A) 350 lbf/in^2
(B) 450 lbf/in^2
(C) 530 lbf/in^2
(D) 690 lbf/in^2

104. Design requirements specify that a cell phone's plastic shell must survive a 6 ft fall onto concrete. Static compressions tests indicate shell fracture at 1250 lbf and 0.005 in deflection. If a factor of safety of 6 is used, what is most nearly the maximum allowable phone weight?

(A) 0.51 lbf
(B) 0.78 lbf
(C) 0.95 lbf
(D) 1.2 lbf

105. A cam follower for a specific application must have zero acceleration at the start of rise. Which cam profile best meets this requirement?

(A) harmonic
(B) cycloidal
(C) parabolic
(D) velocity derivative

106. A long, thin cantilever beam with a rectangular cross section is subject to lateral vibration. If the thickness of the beam is doubled, the beam's fundamental natural frequency will be

(A) divided by four
(B) divided by two
(C) multiplied by two
(D) multiplied by four

107. European Union directives require many products sold in Europe to display the CE marking. What is this mark?

(A) approval mark issued by a nongovernmental certification body
(B) manufacturer's self declaration of conformance
(C) government certification signifying successful product testing
(D) mark of origin indicating a product made in the European Community

108. The moment at support A is most nearly

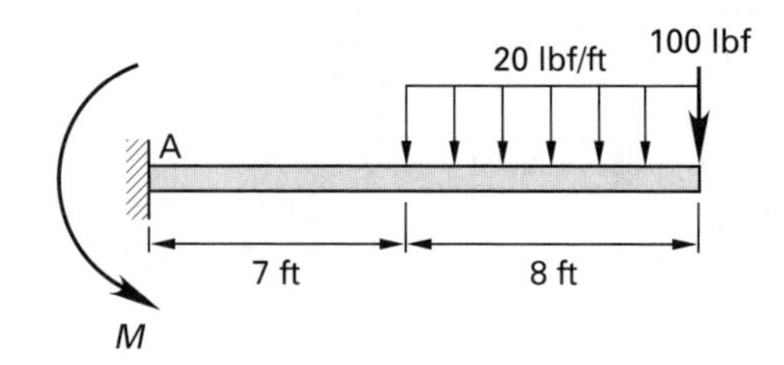

(A) 1900 lbf
(B) 2600 lbf
(C) 3300 lbf
(D) 3900 lbf

109. A 50 lbm mass is connected to the top of a vertical steel rod with a length of 36 in and a diameter of 2 in. The rod is clamped securely at its base. An integral damping system is modeled as a damping coefficient of 1.0 lbf-sec/in. Neglecting the rod's mass, the damping ratio is most nearly

(A) 0.011
(B) 0.036
(C) 0.075
(D) 0.098

110. A tractor pulls a 500 kg crate, initially at rest, up a 26° incline using the cable-pulley system shown. If the tractor moves with a constant velocity of 15 m/s after it is attached to the crate, and the cable remains taut throughout the hauling procedure, what is most nearly the tension in the cable? Neglect the mass of the cable-pulley system, and assume the coefficient of kinetic friction between the crate and ramp is 0.24.

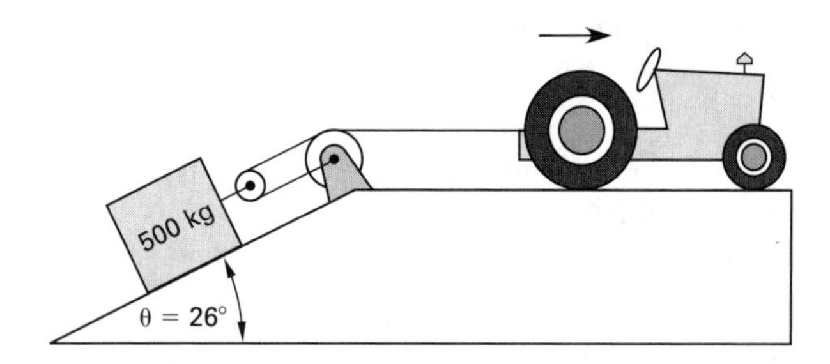

(A) 530 N
(B) 1100 N
(C) 1600 N
(D) 3200 N

111. Zinc blocks attached to the steel hull of a seagoing vessel at various locations below the waterline prevent corrosion. What is the name of this type of corrosion protection?

(A) cathodic protection
(B) anodic protection
(C) passivation
(D) galvanization

112. An 18 lbm mass hangs at the end of a wire wrapped around a solid cylinder that is rotating on a frictionless bearing as shown. The cylinder has a mass of 60 lbm and a radius of 15 in.

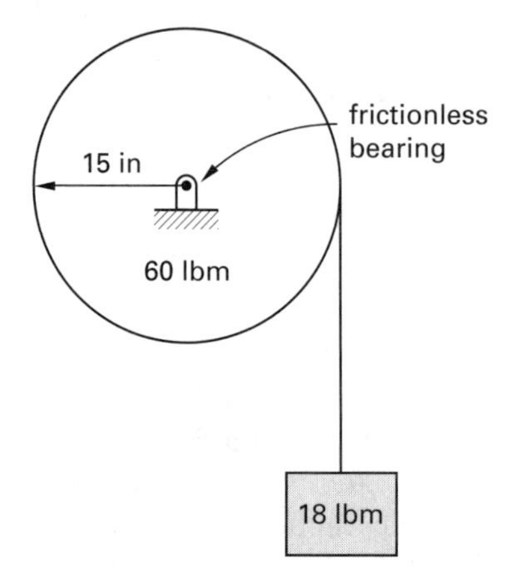

The tension in the string is most nearly

(A) 11 lbf
(B) 13 lbf
(C) 15 lbf
(D) 17 lbf

113. A machine exerts a force that varies cyclically. At the start of operation, the force increases linearly from zero to 32 lbf over 5 sec. It then decreases linearly from 30 lbf to 20 lbf over 4 sec, and then decreases linearly from 20 lbf to zero in 2 sec, at which time the cycle begins again. The average force during the first 20 sec of operation is most nearly

(A) 13 lbf
(B) 15 lbf
(C) 17 lbf
(D) 19 lbf

114. A solid 90 lbm cylindrical wheel with a radius of 5 ft is rotating at 54 rad/sec. The tangential force that must be applied to the wheel's contact surface in order to reduce the rotational speed by one-third in 30 sec is most nearly

(A) 4 lbf
(B) 10 lbf
(C) 40 lbf
(D) 100 lbf

115. Aluminum sheet is fillet welded to create the T-section shown. The weld leg size is 1/8 in, sheet thickness is 1/8 in, the weld length is 10 in, and both sides are welded. The allowable force per inch of weld leg is 5000 lbf/in. The joint is required to withstand an impact load of 1000 lbf suddenly applied parallel to the weld and 2 in from the fixed base as shown. What is most nearly the factor of safety?

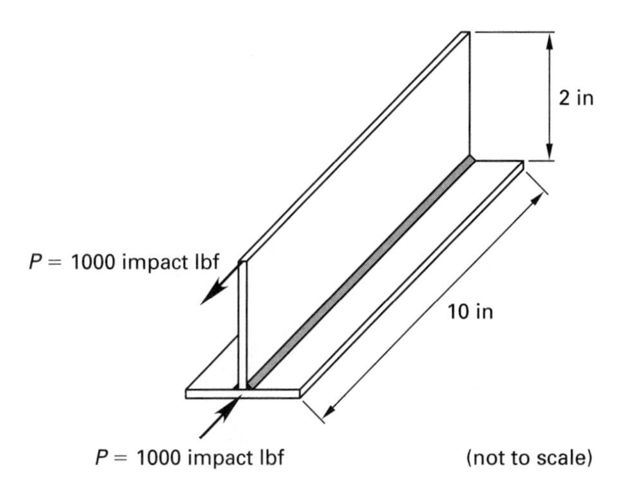

(A) 4.0
(B) 6.0
(C) 8.0
(D) 16.0

116. A joint is made between two metal pieces by closely fitting their surfaces and distributing a molten nonferrous filler metal to the interface by capillary attraction. The pieces to be joined have a melting point of 1400°F, and the filler metal melts at 700°F. This process is most accurately termed

(A) soldering
(B) brazing
(C) welding
(D) forge welding

117. A 0.500 in nominal diameter hole and pin are to be combined during assembly. Drawings indicate tolerances as listed. The basic hole system is used.

parameter	high tolerance (in)	low tolerance (in)
hole diameter	$+1.6 \times 10^{-3}$ in	-0 in
pin diameter	-2.0×10^{-3} in	-3.0×10^{-3} in

This fit is most properly designated as

(A) RC7 free running fit
(B) LC1 locational clearance fit
(C) LT3 locational-transitional fit
(D) FN2 medium drive fit

118. The motion of a lightly damped system is recorded. The amplitudes of two successive cycles of motion are recorded as 0.569 in and 0.462 in. The damping ratio is most nearly

(A) 0.012
(B) 0.033
(C) 0.068
(D) 0.095

119. A truss is constructed of pin-connected rigid members. The force in member BC is most nearly

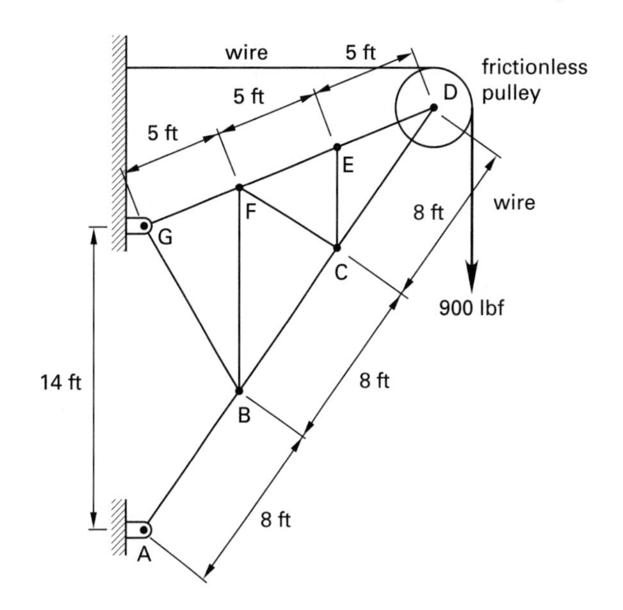

(A) 830 lbf
(B) 880 lbf
(C) 920 lbf
(D) 970 lbf

120. Three parts stacked in assembly have individual nominal dimensions and tolerances as listed. Tolerances are normally distributed and represent $\pm 3\sigma$ from the mean value.

part	dimension (in)
1	0.8 ± 0.1
2	1.0 ± 0.2
3	1.2 ± 0.3

The toleranced height of the stack assembly is most nearly

(A) 3.0 ± 0.06 in
(B) 3.0 ± 0.09 in
(C) 3.0 ± 0.4 in
(D) 3.0 ± 0.6 in

STOP!

DO NOT CONTINUE!

This concludes the Afternoon Session of the examination. If you finish early, check your work and make sure that you have followed all instructions. After checking your answers, you may turn in your examination booklet and answer sheet and leave the examination room. Once you leave, you will not be permitted to return to work or change your answers.

Afternoon Session
Thermal and Fluids Systems

121. An irrigation system is being designed in which a water tower will supply the irrigation ditches. The water is to be discharged at 60 gpm. The gate valve that will control the water discharge is to operate fully open, and the pressure drop across this valve is to be no more than 5 psi. Of the following choices, the smallest suitable size for the gate valve is

(A) $^1/_2$ in
(B) $^3/_4$ in
(C) 1 in
(D) $1^1/_4$ in

122. Thirty-two balls are contained in a 2 ft wide, 2 ft deep, 3 ft tall container for a lottery drawing. A fan blowing 60°F air upward from beneath the container forces the balls upward in a clear tube so as to display the number on each ball. If the balls each have a diameter of 1.5 in and weigh 0.00516 lbf, the flow rate through the container must be at least

(A) 4500 cfm
(B) 7200 cfm
(C) 11,000 cfm
(D) 22,000 cfm

123. The power output from a generator that is driven by an impulse turbine is 122 000 kW. The turbine wheel, which is mounted on a 3 ft diameter steel shaft, has a 20 ft diameter and is spun by a 2 ft diameter water jet. Assuming ideal conditions ($v_{jet} = 2v_{vane}$ and the jet is deflected through 180°), the angular speed of the turbine wheel is most nearly

(A) 15 rpm
(B) 150 rpm
(C) 1300 rpm
(D) 2600 rpm

124. In the system shown, the head loss (in feet) in the pipes is given by $0.015(L/D)(v^2/2g)$, where L is the length of pipe in feet, D is the pipe diameter in feet, v is the velocity in ft/sec, and g is the acceleration of gravity in ft/sec^2. The maximum and minimum pressures in the system are most nearly

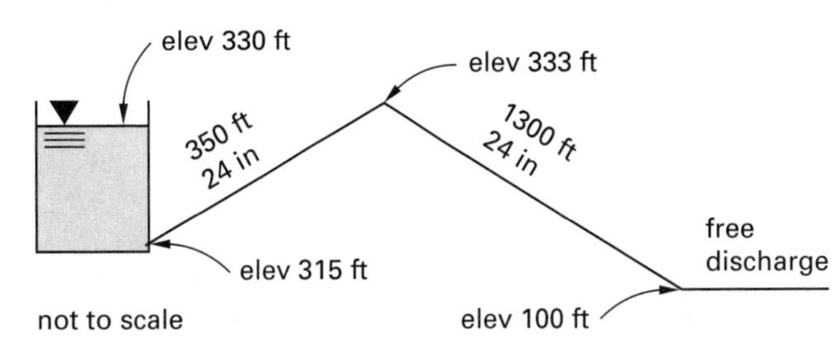

(A) 6.5 psi, −28 psi
(B) 39 psi, −8.0 psi
(C) 93 psi, 7.8 psi
(D) 100 psi, 0.0 psi

125. In the system shown, the flow rate of 68°F water is 1500 gpm, and the head loss (in feet) in the pipes is given by $0.02(L/D)(v^2/2g)$, where L is the length of pipe in feet, D is the pipe diameter in feet, v is the velocity in ft/sec, and g is the acceleration of gravity in ft/sec^2. If the motor driving the pump is 90% efficient and the pump is 80% efficient, the power that must be supplied to the motor is most nearly

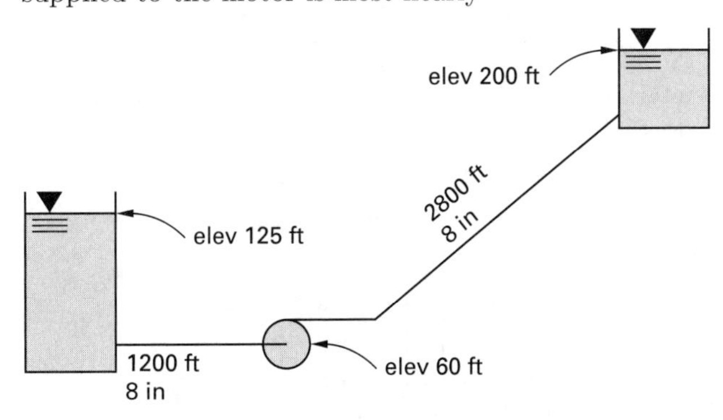

(A) 52 kW
(B) 69 kW
(C) 97 kW
(D) 130 kW

126. A centrifugal pump with the pump curve given is to be installed as shown. The line uses schedule-40 steel pipe and contains two long-radius 90° elbows and one gate valve. The pipe friction factor is 0.015. The discharge will be most nearly

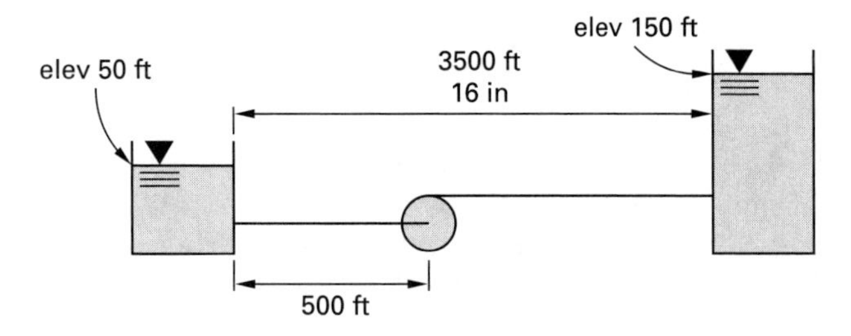

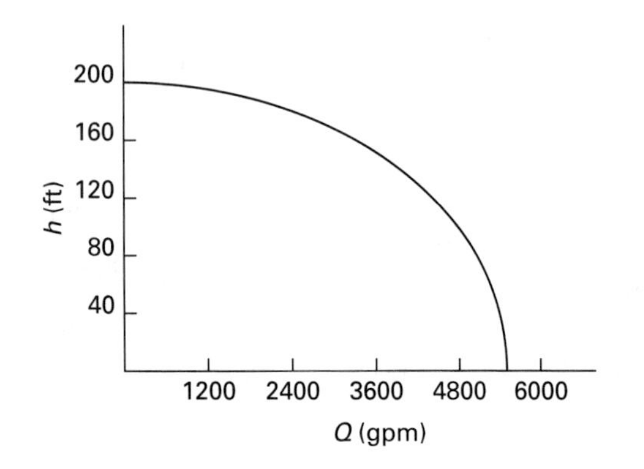

(A) 2400 gpm
(B) 3000 gpm
(C) 4200 gpm
(D) 5100 gpm

127. An ideal compressible fluid is flowing from point A to point B. The properties of the fluid at points A and B are as follows.

	A	B
density	0.381 lbm/ft^3	0.0508 lbm/ft^3
internal energy	119.1 Btu/lbm	53.1 Btu/lbm
pressure	98.4 psia	5.85 psia

The change in enthalpy between points A and B is most nearly

(A) −93 Btu/lbm (decrease)
(B) −66 Btu/lbm (decrease)
(C) −27 Btu/lbm (decrease)
(D) 77 Btu/lbm (increase)

128. A 3 in diameter steel sphere achieves a terminal velocity of 122 in/sec when dropping through a tall column of liquid. The density of the liquid is 87 lbm/ft^3. The density of steel is 488 lbm/ft^3. The final drag coefficient of the sphere is most nearly

(A) 0.22
(B) 0.34
(C) 0.40
(D) 0.48

129. Water in a supply tank is used to form a small fountain of water as shown. Forty feet of 4 in schedule-40 steel pipe containing five long-radius 90° elbows and one gate valve run from the bottom of the tank to the discharge point. The opening in the tank has a sharp edge. If the minimum height that the 50°F water must achieve is 4 ft, the water depth in the tank should never be less than

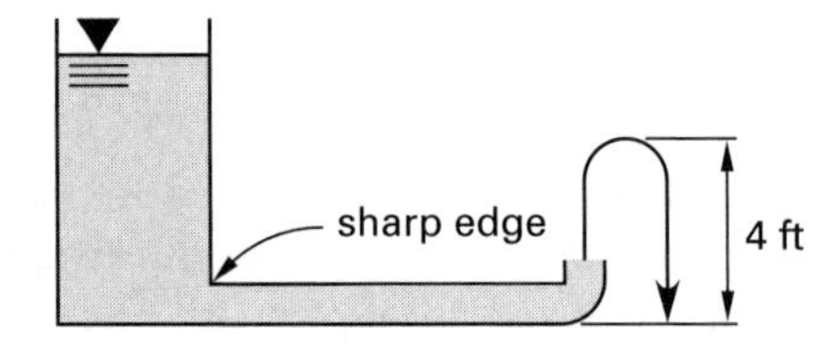

(A) 15 ft
(B) 21 ft
(C) 33 ft
(D) 54 ft

130. A feedwater pump delivers 100 gpm water to a system. Because of concerns regarding the potential for erosion at the elbows, it has been specified that the water velocity in the pipe must not exceed 7.0 ft/sec. What minimum size schedule-40 pipe (nominal) will carry the flow without exceeding the maximum velocity criteria?

(A) 2 in
(B) 2$^1/_2$ in
(C) 4 in
(D) 20 in

131. A two-cylinder Otto cycle internal combustion engine produces 33 hp at 400 rpm. The bore is 8 in, and the stroke is 8 in. What is most nearly the brake mean effective pressure?

(A) 3.4 psig
(B) 21 psig
(C) 41 psig
(D) 81 psig

132. A hydraulic wood splitter with a double-acting cylinder operating at 2500 psi has a 24 in stroke in both directions. The hydraulic oil at normal operating temperature has a specific gravity of 0.85. The wood splitter has a 3 in bore and a 1 in diameter cylinder rod. The maximum retraction time for the cylinder rod is 5 sec. A four-way valve with a valve flow coefficient of 2.9 is used for cycling the wood splitter. The pressure drop across the four-way valve is most nearly

(A) 6.2 psi
(B) 6.7 psi
(C) 7.3 psi
(D) 7.9 psi

133. A heat pump supplies 900×10^3 Btu/day to a 3500 ft^2 home in order to maintain the temperature inside at 78°F when the temperature outside is 30°F. Electricity costs 3.7 cents/kW-hr. The minimum cost per 30 day month to operate the heat pump is most nearly

(A) $18
(B) $26
(C) $41
(D) $89

134. Enriched uranium (with a thermal conductivity of 20 Btu/hr-ft-°F) fuel plates 0.25 in thick are used in a nuclear reactor core. The coolant maintains one side of a plate at the edge of the core at 500°F and the other side at 750°F. If the internal heat generation of the plate is a uniform 9×10^7 Btu/hr-ft^3, the maximum temperature in the plate is most nearly

(A) 580°F
(B) 890°F
(C) 1700°F
(D) 1800°F

135. A counterflow shell-and-tube heat exchanger is to be designed using 1 in OD, 0.875 in ID, 30 ft long tubing. 40,500 lbm/hr of water at 55°F are used to cool 45,000 lbm/hr of a 95% ethyl alcohol solution (with a specific heat of 0.9 Btu/lbm-°F) from 160°F to 110°F. If the overall coefficient of heat transfer based on the outer tube area is 75 Btu/hr-ft^2-°F, then the number of tubes in the heat exchanger is most nearly

(A) 33
(B) 69
(C) 82
(D) 130

136. Steam at 900°F enters a turbine that operates in a Rankine cycle. If the cycle operates between pressures of 1000 psia and 20 psia, the efficiency of the cycle is most nearly

(A) 10%
(B) 20%
(C) 30%
(D) 50%

137. Four pounds per minute of steam at 200 psia and 600°F are mixed with 11 lbm/min of steam at 200 psia and 70% quality in an adiabatic steady-flow device. The quality of the outlet mixture is most nearly

(A) 71%
(B) 82%
(C) 91%
(D) 97%

138. The rectangular wall of a furnace consists of 3 in fire-clay brick surrounded by 0.25 in of steel on the outside. There are six 0.25 in diameter mild steel bolts per square foot connecting the steel to the brick and extending completely through both the steel and the brick. The furnace is surrounded by 70°F air (with a film coefficient of 1.65 Btu/hr-ft^2-°F), while the inner surface of the brick is maintained at a constant 1000°F. The heat flux per square foot through this wall is most nearly

(A) 160 Btu/hr
(B) 300 Btu/hr
(C) 450 Btu/hr
(D) 930 Btu/hr

139. A residential home heating demand of 110,000 Btu/hr is supplied by a heat pump system that uses compressed air to heat water circulating through the house. The water enters the house at 120°F and re-enters the exchanger at 92°F. The air enters the exchanger at 190°F and leaves at 125°F. The heat exchanger is a one-shell pass and two-tube pass counterflow exchanger with an overall coefficient of heat transfer of 35 Btu/hr-ft^2-°F and an area-to-volume ratio of 95 ft^2/ft^3. For this heat exchanger, the number of transfer units (NTU) is most nearly

(A) 1.7
(B) 2.5
(C) 3.6
(D) 4.2

140. A 200 ton hydraulic press has a 90% efficient, 1200 rpm electric motor that is direct coupled to an 85% efficient hydraulic pump. The motor has a power factor of 0.85. The pump has a rated displacement of 1.93. The system's hydraulic pressure is 2500 psig. The press is powered by a 208 V three-phase electrical service. The current drawn by the motor is most nearly

(A) 30 A
(B) 50 A
(C) 60 A
(D) 80 A

141. A 40 W incandescent light bulb emitting 1500 lm has a surface temperature of 260°F. Still air at 80°F surrounds the bulb, and the bulb is modeled as a 2 in diameter sphere. The heat lost to free convection is most nearly

(A) 0.50 W
(B) 6.9 W
(C) 16 W
(D) 24 W

142. The supply side of a closed-loop geothermal home heating system is plumbed with 400 ft of 1 in schedule-40 PVC pipe. The pipe is buried 3 ft underground and runs from the house to the bottom of a pond. The portion of the pipe that is underground is insulated with a 0.125 in thick layer of insulation ($k = 0.03$ Btu/hr-ft-°F). The glycol solution traveling through the 250 ft of pipe on the bottom of the pond is heated to 56°F, and the ground surface temperature is 20°F. The maximum heat loss per unit length of pipe experienced anywhere in the loop is most nearly

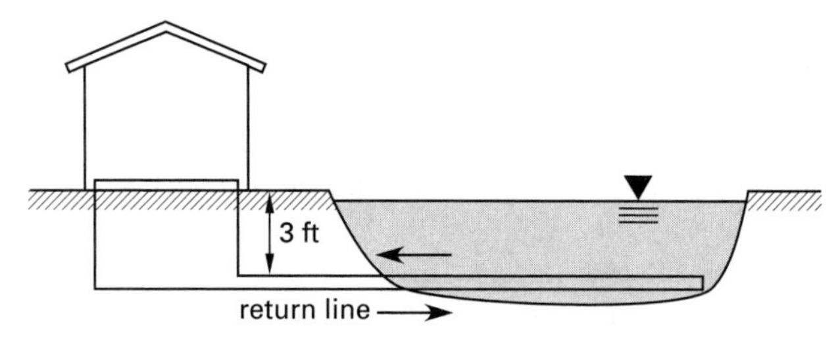

(A) 0.5 Btu/hr-ft
(B) 1.5 Btu/hr-ft
(C) 13 Btu/hr-ft
(D) 27 Btu/hr-ft

143. A 240 V electric jockey pump adds 300 ft of pressure head to a 60°F, 1.5 lbm/sec water flow. At operating speed, the motor is 95% efficient, and the pump is 65% efficient. What is most nearly the motor current draw?

(A) 2.5 A
(B) 4.1 A
(C) 6.2 A
(D) 11.0 A

144. A steel rod with a circular cross section has both ends built in between two walls. The rod is 11.5 ft long and has a diameter of 2.4 in. If the temperature is gradually increased, the rise in temperature at which the rod will first begin to buckle is most nearly

(A) 68°F
(B) 102°F
(C) 136°F
(D) 170°F

145. The thermocouple reading of a flame in a furnace is 550°F, and the temperature of the inside furnace walls is 400°F. The convection coefficient inside the furnace is 25 Btu/hr-ft^2-°F. The true temperature of the flame is most nearly

(A) 580°F
(B) 620°F
(C) 850°F
(D) 1000°F

146. A two-stage air compressor with an ideal intercooler takes air at 14.7 psia and 70°F and compresses it to 200 psia. The pressure staging is ideal. The mass flow rate of air is 10 lbm/min. The heat removed by the intercooler is most nearly

(A) 570 Btu/min
(B) 870 Btu/min
(C) 2400 Btu/min
(D) 4500 Btu/min

147. Water exits from an open holding tank to the atmosphere through a small opening 11 ft below the surface of the water. The volume of water in the tank remains constant. The ideal velocity of the water leaving the tank is most nearly

(A) 4.7 ft/sec
(B) 19 ft/sec
(C) 27 ft/sec
(D) 88 ft/sec

148. Atmospheric air flowing at 800 ft^3/min and at a wet-bulb temperature of 75°F and a dry-bulb temperature of 98°F is conditioned to a dry-bulb temperature of 67°F and 50% relative humidity in an evaporative cooler. The amount of heat removed by the cooling system and added by the heating system is most nearly

(A) heat removed = 1100 Btu/hr
heat added = 260 Btu/hr
(B) heat removed = 1600 Btu/hr
heat added = 3700 Btu/hr
(C) heat removed = 49,000 Btu/hr
heat added = 24,000 Btu/hr
(D) heat removed = 65,000 Btu/hr
heat added = 15,000 Btu/hr

149. Steam at 250 psia enters an isentropic turbine operating in a Rankine cycle. The steam exits the turbine at 100°F. The moisture content of the steam in the turbine is not to exceed 8%. The minimum temperature to which the high-pressure steam must be heated is most nearly

(A) 170°F
(B) 530°F
(C) 1000°F
(D) 1320°F

150. At the beginning of the compression stroke of an air-standard Otto cycle, the pressure is 14.7 psia and the temperature is 68°F. The maximum pressure in the cycle is 510 psia, and the compression ratio is 8.5:1. The difference in efficiency between this cycle and a Carnot cycle operating between the same temperatures is most nearly

(A) 10%
(B) 20%
(C) 30%
(D) 40%

151. The net output power from a steam-based Rankine cycle is 250 MW. The cycle has steam entering an isentropic turbine at 1100°F and 1250 psia, and cooling water entering the condenser at 56°F and leaving as saturated liquid at 93°F. The condenser pressure is 5 psia. The mass flow rate of steam is most nearly

(A) 2.9×10^4 lbm/hr
(B) 3.3×10^5 lbm/hr
(C) 4.7×10^5 lbm/hr
(D) 1.6×10^6 lbm/hr

152. A pressure control valve for limiting hydraulic system pressure is redesigned to use a helical spring with squared and ground ends. The spring will be made from 0.10 in diameter wire having a shear modulus of 11.5×10^6 psi. The maximum system operating pressure is to be 250 psi. The effective control area where the spring ultimately acts to limit system pressure is 0.123 in^2. The maximum spring deflection is to be 7/16 in. The spring index is 8. The spring must have

(A) 2 coils
(B) 4 coils
(C) 6 coils
(D) 8 coils

153. A regenerative vapor power cycle with one open feedwater heater operates between pressures of 600 psia and 10 psia with a turbine inlet steam temperature of 860°F. Steam at 130 psia is extracted and sent to the open feedwater heater. The percentage of the total flow passing through the second-stage isentropic turbine is most nearly

(A) 14%
(B) 51%
(C) 75%
(D) 86%

154. Two metal blocks are brought into contact with each other and allowed to reach thermal equilibrium. One has a mass of 60 lbm, is made of copper, and is at 50°F. The other has a mass of 30 lbm, is made of stainless steel, and is at 420°F. If the blocks are considered as a system and losses to the surroundings are neglected, the final temperature is most nearly

(A) 170°F
(B) 190°F
(C) 230°F
(D) 260°F

155. A gas pipeline is being added near a small residential neighborhood just outside of a town. The proposed pipeline is 1 mi long and will carry gas at 30 psig and 300°F. There are 25 homes within 1/8 mi of the proposed pipeline, all homes are one or two stories, and there are no plans for significant further development. What ASME location class is most appropriate for this area?

(A) Class 1
(B) Class 2
(C) Class 3
(D) Class 4

156. Water flows through a cast-iron pipe at 1800 gpm. The inside diameter of the pipe is 10 in. The coefficient of friction is 0.0195. The pressure drop over an 185 ft length of pipe is most nearly

(A) 1.6 psi
(B) 3.6 psi
(C) 16 psi
(D) 19 psi

157. A natural gas has the following molar analysis.

CH_4	80.32%
C_2H_6	5.75%
C_3H_8	1.79%
C_4H_{10}	1.69%
N_2	10.45%

When the gas is burned in air, the following dry analysis describes the products.

CO_2	7.9%
CO	0.1%
O_2	7%
N_2	85%

The air/fuel ratio is most nearly

(A) 4.8 kmol air/kmol fuel
(B) 7.2 kmol air/kmol fuel
(C) 11 kmol air/kmol fuel
(D) 14 kmol air/kmol fuel

158. A six-cylinder, four-stroke engine has a 100.4 mm bore and a 84.4 mm stroke. The brake mean effective pressure while running at 4200 rpm is 120 psig. The brake horsepower is most nearly

(A) 100 hp
(B) 160 hp
(C) 180 hp
(D) 210 hp

159. A 110 hp internal combustion engine burns propane at a rate of 0.01 lbm/sec when mixed with theoretical air. The engine takes in air and fuel at 77°F and 1 atm. The air-fuel mixture burns completely and leaves the engine at 1240°F. The rate of heat transfer from the engine and exhaust is most nearly

(A) −140 Btu/sec
(B) −93 Btu/sec
(C) −68 Btu/sec
(D) 33 Btu/sec

160. Octane is burned completely with theoretical air at atmospheric pressure. The dew point of the products is most nearly

(A) 130°F
(B) 150°F
(C) 200°F
(D) 210°F

STOP!

DO NOT CONTINUE!

This concludes the Afternoon Session of the examination. If you finish early, check your work and make sure that you have followed all instructions. After checking your answers, you may turn in your examination booklet and answer sheet and leave the examination room. Once you leave, you will not be permitted to return to work or change your answers.

Answer Key

Morning Session

Question	Answer	Question	Answer	Question	Answer	Question	Answer
1.	C	11.	B	21.	B	31.	B
2.	D	12.	A	22.	B	32.	B
3.	C	13.	C	23.	D	33.	C
4.	B	14.	A	24.	C	34.	B
5.	B	15.	D	25.	D	35.	A
6.	D	16.	C	26.	A	36.	D
7.	C	17.	D	27.	A	37.	C
8.	B	18.	C	28.	D	38.	A
9.	C	19.	D	29.	D	39.	C
10.	B	20.	C	30.	C	40.	B

Afternoon Session—HVAC and Refrigeration

No.	A	B	C	D
41.	●	B	C	D
42.	A	●	C	D
43.	●	B	C	D
44.	●	B	C	D
45.	A	B	C	●
46.	●	B	C	D
47.	A	B	●	D
48.	A	B	●	D
49.	A	●	C	D
50.	A	B	●	D
51.	A	B	●	D
52.	A	B	C	●
53.	A	B	●	D
54.	A	●	C	D
55.	A	B	C	●
56.	A	B	C	●
57.	A	B	C	●
58.	A	●	C	D
59.	A	B	●	D
60.	A	B	●	D
61.	A	●	C	D
62.	A	B	●	D
63.	A	B	●	D
64.	A	B	●	D
65.	A	B	●	D
66.	A	●	C	D
67.	A	B	C	●
68.	A	B	C	●
69.	A	B	●	D
70.	●	B	C	D
71.	A	●	C	D
72.	A	●	C	D
73.	A	●	C	D
74.	●	B	C	D
75.	A	B	●	D
76.	A	B	●	D
77.	A	B	C	●
78.	●	B	C	D
79.	A	B	●	D
80.	A	●	C	D

Afternoon Session—Machine Design

No.	A	B	C	D
81.	A	●	C	D
82.	A	B	●	D
83.	A	B	C	●
84.	A	B	●	D
85.	●	B	C	D
86.	A	●	C	D
87.	●	B	C	D
88.	A	B	●	D
89.	A	B	●	D
90.	●	B	C	D
91.	A	B	●	D
92.	A	B	C	●
93.	A	●	C	D
94.	●	B	C	D
95.	A	●	C	D
96.	A	B	C	●
97.	●	B	C	D
98.	A	B	●	D
99.	A	●	C	D
100.	A	●	C	D
101.	●	B	C	D
102.	A	B	●	D
103.	●	B	C	D
104.	A	B	C	●
105.	A	●	C	D
106.	A	B	●	D
107.	A	●	C	D
108.	A	B	●	D
109.	A	●	C	D
110.	A	B	●	D
111.	●	B	C	D
112.	●	B	C	D
113.	A	B	C	●
114.	●	B	C	D
115.	A	B	●	D
116.	●	B	C	D
117.	●	B	C	D
118.	A	●	C	D
119.	A	B	●	D
120.	A	B	●	D

Afternoon Session—Thermal and Fluids Systems

No.	A	B	C	D
121.	A	●	C	D
122.	A	●	C	D
123.	A	●	C	D
124.	●	B	C	D
125.	A	B	●	D
126.	A	B	●	D
127.	●	B	C	D
128.	A	B	C	●
129.	A	●	C	D
130.	A	●	C	D
131.	A	B	C	●
132.	●	B	C	D
133.	A	●	C	D
134.	A	●	C	D
135.	A	●	C	D
136.	A	B	●	D
137.	A	●	C	D
138.	A	B	C	●
139.	●	B	C	D
140.	A	●	C	D
141.	A	●	C	D
142.	●	B	C	D
143.	A	●	C	D
144.	●	B	C	D
145.	●	B	C	D
146.	●	B	C	D
147.	A	B	●	D
148.	A	B	C	●
149.	A	B	●	D
150.	A	●	C	D
151.	A	B	C	●
152.	A	B	●	D
153.	A	B	C	●
154.	A	●	C	D
155.	A	●	C	D
156.	●	B	C	D
157.	A	B	C	●
158.	A	●	C	D
159.	A	B	●	D
160.	●	B	C	D

Solutions
Morning Session

1. PVC, polystyrene, and ABS are examples of amorphous polymers, while polypropylene is a crystalline polymer. In general, crystalline polymers have better chemical resistance than amorphous polymers.

The answer is (C).

2. The truss is pin-connected, and member EF is the only vertical member framing into joint F, so EF is a zero-force member. If member EF were loaded, either in tension or compression, the vertical component of that load could not be transferred out of the joint.

The answer is (D).

3. The total dynamic head provided by the pump, h_{pump}, is that required to overcome the pressure of the pressurized tank, the friction and fitting losses (h_{losses}), and the static discharge head to the pressurized tank waterline, less the positive static head of the water column to the storage tank above.

Bernoulli's equation describes the energy (as head) relationships as

$$h_{\text{pump}} + \frac{p_1}{\gamma_1} + \frac{\text{v}_1^2}{2g} + z_1 = \frac{p_2}{\gamma_2} + \frac{\text{v}_2^2}{2g} + z_2 + h_{\text{losses}}$$

The equation can be rearranged to solve for the required pump head, choosing the water surface in the open tank as point 1 and the water surface of the pressurized tank as point 2. The velocities v_1 and v_2 are approximately zero.

$$\begin{aligned} h_{\text{pump}} &= \frac{p_2 - p_1}{\gamma} + \frac{\text{v}_2^2 - \text{v}_1^2}{2g} + z_2 - z_1 + h_{\text{losses}} \\ &= \frac{\left(15.0\ \frac{\text{lbf}}{\text{in}^2} - 0\ \frac{\text{lbf}}{\text{in}^2}\right)\left(12\ \frac{\text{in}}{\text{ft}}\right)^2}{62.4\ \frac{\text{lbf}}{\text{ft}^3}} \\ &\quad + 7\text{ ft water} - 15\text{ ft water} + 12\text{ ft water} \\ &= 38.65\text{ ft water} \quad (39\text{ ft water}) \end{aligned}$$

The answer is (C).

4. In a principal, cavalier, or sectional view, at least one projection (also called a projection plane) is parallel with the plane of the screen or paper (or in other words, perpendicular to the direction from which the object is being viewed). The view shown is isometric.

The answer is (B).

5. The heat added by the compressor comes from the work done by the compressor. The COP of the process was given as 4.4. The COP is the refrigeration effect divided by the power.

$$\text{COP} = \frac{\dot{q}_{\text{entering}}}{P_{\text{in}}}$$

$$P_{\text{in}} = \frac{\dot{q}_{\text{entering}}}{\text{COP}}$$

The capacity required by the evaporator to cool the water is

$$\begin{aligned} \dot{q}_{\text{evaporator}} &= \dot{m}c_p\Delta T \\ &= \dot{V}\rho c_p\left(T_{\text{entering}} - T_{\text{leaving}}\right) \\ &= \left(640\ \frac{\text{gal}}{\text{min}}\right)\left(\frac{1\text{ ft}^3}{7.48\text{ gal}}\right)\left(60\ \frac{\text{min}}{\text{hr}}\right) \\ &\quad \times \left(62.4\ \frac{\text{lbm}}{\text{ft}^3}\right)\left(1.0\ \frac{\text{Btu}}{\text{lbm-°F}}\right) \\ &\quad \times (55°\text{F} - 43°\text{F})\left(\frac{1\text{ ton}}{12{,}000\ \frac{\text{Btu}}{\text{hr}}}\right) \\ &= 320\text{ tons} \end{aligned}$$

The compressor power is

$$\begin{aligned} P_{\text{in}} &= \frac{320\text{ tons}}{4.4} \\ &= 72.7\text{ tons} \quad (73\text{ tons}) \end{aligned}$$

The answer is (B).

6. The reciprocating saw develops 600 W (0.6 kW) of power, which the inverter must supply. The inverter has a 68% efficiency, so the inverter power is

$$P_{\text{inv}} = \frac{P_{\text{saw}}}{\eta} = \frac{600\text{ W}}{0.68} = 882.4\text{ W}$$

Method 1:

Power equals current times voltage ($P = IV$), so the inverter current is

$$I_{\text{inv}} = \frac{P_{\text{inv}}}{V_{\text{inv}}} = \frac{882.4\ \text{W}}{12\ \text{V}} = 73.5\ \text{A} \quad (74\ \text{A})$$

Method 2:

$$P_{\text{inv}} = I_{\text{inv}} V_{\text{inv}} = I_{\text{inv}}(I_{\text{inv}} R_{\text{inv}}) = I^2_{\text{inv}} R_{\text{inv}}$$

$$\begin{aligned} I_{\text{inv}} &= \sqrt{\frac{P_{\text{inv}}}{R_{\text{inv}}}} = \sqrt{\frac{882.4\ \text{W}}{0.16\ \Omega}} \\ &= 74.3\ \text{A} \quad (74\ \text{A}) \end{aligned}$$

The answer is (D).

7. Because the relief valve discharges into atmospheric pressure and the hydrostatic head is present in both the pump and relief lines, the total pressure drop across the valve is the oil pressure.

Valve flow rate is related to the pressure drop across the valve through the following formula (which is not dimensionally consistent). C_v is the valve flow coefficient.

$$\begin{aligned} Q_{\text{gpm}} &= C_v \sqrt{\frac{\Delta p_{\text{psi}}}{\text{SG}}} \\ &= 0.53 \sqrt{\frac{5.775\ \frac{\text{lbf}}{\text{in}^2}}{0.9}} \\ &= 1.34 \quad (1.3\ \text{gpm}) \end{aligned}$$

The answer is (C).

8. The types of joints in the four answer choices are shown here.

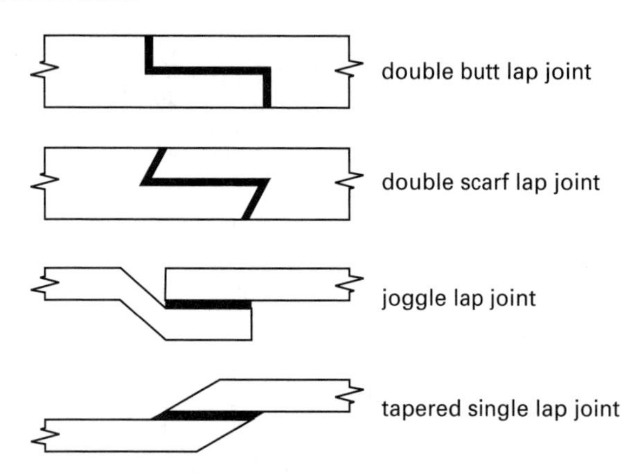

The joggle lap joint (C) can be formed with simple bending operations, without machining. Like a simple lap joint (not shown), the joggle lap joint is relatively subject to peel and cleavage failures. The tapered single lap joint (D), double butt lap joint (A), and double scarf lap joint (B) offer progressively better performance in applications with bending.

The answer is (B).

9. A liquid's kinematic viscosity in Saybolt universal seconds is the time it takes for 60 cm^3 of the liquid to flow through a calibrated tube at a controlled temperature. Intuitively, then, neither 0.9 SUS nor 32,000 SUS is a plausible answer choice for SAE 10W-30 oil.

For viscosities greater than about 240 SUS, one centistoke is equal to about 0.216 SUS. The kinematic viscosity in Saybolt universal seconds is

$$\nu = (110\ \text{cSt})\left(\frac{1\ \text{cSt}}{0.216\ \text{SUS}}\right) = 509\ \text{SUS} \quad (500\ \text{SUS})$$

The answer is (C).

10. The reheat load is the rate of sensible heat required to bring the air leaving the coil to the conditions required to supply the gallery room.

$$\begin{aligned} \dot{q}_{\text{reheat}} &= \dot{m} c_p (T_{\text{leaving heating coil}} - T_{\text{entering heating coil}}) \\ &= \left(8000\ \frac{\text{ft}^3}{\text{min}}\right)\left(60\ \frac{\text{min}}{\text{hr}}\right)\left(0.075\ \frac{\text{lbm}}{\text{ft}^3}\right) \\ &\quad \times \left(0.24\ \frac{\text{Btu}}{\text{lbm-°F}}\right)(60°\text{F} - 53°\text{F}) \\ &= 60{,}480\ \text{Btu/hr} \quad (60{,}000\ \text{Btu/hr}) \end{aligned}$$

The answer is (B).

11. At $T = 40°\text{C}$, the saturation pressure of water vapor is 7.38 kPa.

Determine the specific humidity of the air at the inlet and outlet. The specific humidity, or mass ratio of water vapor to dry air, is given by

$$\omega = \frac{m_w}{m_a}$$

The specific humidity is expressed in units kg/kg or as a dimensionless quantity. Under the ideal gas law, the specific humidity can be written as follows.

$$\begin{aligned} \omega &= \frac{m_w}{m_a} = \frac{\frac{p_w V}{R_w T}}{\frac{p_a V}{R_a T}} \\ &= \frac{0.622 p_w}{p_a} \\ &= \frac{0.622 p_w}{p - p_w} \end{aligned}$$

Initially the air is dry, so $\omega_1 = 0$. The air exiting the channel at point 3 is saturated at 40°C. Therefore, the specific humidity at point 3 is

$$\begin{aligned} \omega_3 &= \frac{(0.622)(7.38\ \text{kPa})}{100\ \text{kPa} - 7.38\ \text{kPa}} \\ &= 0.049 \end{aligned}$$

Since the channel is insulated, it can be treated as a two-inlet, one-outlet adiabatic, steady-flow system. Apply conservation of mass to both the dry air flow and water vapor flow through the channel. The mass of air entering equals the mass leaving.

$$\dot{m}_{a,1} = \dot{m}_{a,3}$$

The increase in water vapor is due to evaporation from the pool.

$$\dot{m}_{w,1} + \dot{m}_{\text{evap}} = \dot{m}_{w,3}$$

Combining these relations yields

$$\begin{aligned} \dot{m}_{\text{evap}} &= \dot{m}_{w,3} - \dot{m}_{w1} \\ &= \dot{m}_{a,3}\omega_3 - \dot{m}_{a,1}\omega_1 \\ &= \dot{m}_a(\omega_3 - \omega_1) \\ &= \dot{m}_a\omega_3 \\ &= \left(12\ \frac{\text{kg}}{\text{s}}\right)(0.049) \\ &= 0.588\ \text{kg/s} \quad (0.6\ \text{kg/s}) \end{aligned}$$

The answer is (B).

12. From a saturated water temperature table, the water vapor saturation pressure at temperature $T = 33°\text{C}$ (306K) is

$$p_{\text{sat},33°\text{C}} = 5\ \text{kPa}$$

The partial pressure of the dry air in the room is equal to the difference between the total pressure in the room and the vapor pressure of the moisture in the air. The relative humidity, ϕ, is the ratio of the vapor pressure, p_w, to the saturation pressure of water at a given temperature. That is,

$$\phi = \frac{p_w}{p_{\text{sat}}}$$

The partial pressure of the vapor is

$$\begin{aligned} p_w &= \phi p_{\text{sat}} \\ &= (0.40)(5\ \text{kPa}) \\ &= 2\ \text{kPa} \end{aligned}$$

The answer is (A).

13. From a saturated steam-by-temperature table, at a temperature of 5°C, the saturation pressure of water vapor is

$$p_{\text{sat},5°\text{C}} = 0.872\ \text{kPa}$$

From the same table at 25°C,

$$p_{\text{sat},25°\text{C}} = 3.17\ \text{kPa}$$

Draw a schematic diagram of the air conditioning process.

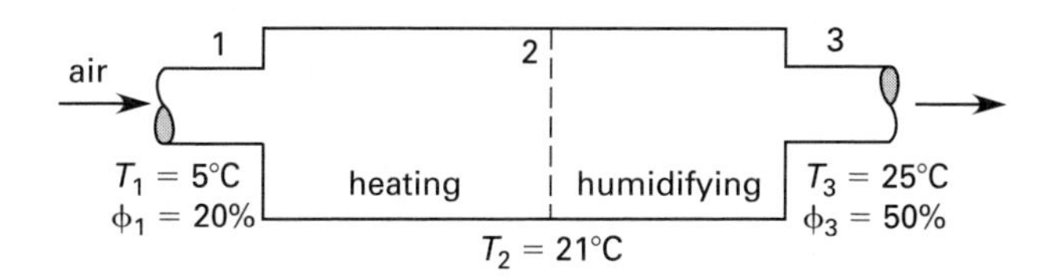

Solve for the specific humidity of the air (in units kilograms of water vapor per kilogram dry air, denoted kg/kg or expressed as a dimensionless quantity).

Since the specific humidity does not change in the heating process, $\omega_2 = \omega_1$.

$$\begin{aligned} \omega_2 &= \omega_1 \\ &= \frac{0.622\phi_1 p_{\text{sat},1}}{p - \phi_1 p_{\text{sat},1}} \\ &= \frac{(0.622)(0.20)(0.872\ \text{kPa})}{100\ \text{kPa} - (0.20)(0.872\ \text{kPa})} \\ &= 0.00109 \end{aligned}$$

At the outlet,

$$\begin{aligned} \omega_3 &= \frac{0.622\phi_3 p_{\text{sat},3}}{p - \phi_3 p_{\text{sat},3}} \\ &= \frac{(0.622)(0.50)(3.17\ \text{kPa})}{100\ \text{kPa} - (0.50)(3.17\ \text{kPa})} \\ &= 0.0100 \end{aligned}$$

For the dry air in the airflow,

$$\begin{aligned} \dot{m}_{a,3} &= \dot{m}_{a,2} = \dot{m}_{a,1} \\ &= \dot{V}_1\rho_1 \\ &= \dot{m}_a \end{aligned}$$

By the ideal gas law,

$$\begin{aligned} \rho_1 &= \frac{p_1}{RT_1} = \frac{100\ \text{kPa}}{\left(0.287\ \frac{\text{kJ}}{\text{kg}\cdot\text{K}}\right)(278\text{K})} \\ &= 1.25\ \text{kg/m}^3 \end{aligned}$$

Therefore,

$$\begin{aligned} \dot{m}_a &= \dot{V}_1\rho_1 = \left(1\ \frac{\text{m}^3}{\text{s}}\right)\left(1.25\ \frac{\text{kg}}{\text{m}^3}\right) \\ &= 1.25\ \text{kg/s} \end{aligned}$$

The hot steam contributes to the final vapor content.

$$\dot{m}_{w,2} + \dot{m}_{\text{hs}} = \dot{m}_{w,3}$$

Solving for the hot steam contribution yields

$$\dot{m}_{\text{hs}} = \dot{m}_{w,3} - \dot{m}_{w,2}$$

By definition,

$$\omega = \frac{\dot{m}_w}{\dot{m}_a}$$

Therefore,

$$\begin{aligned}\dot{m}_{\text{hs}} &= \omega_3\dot{m}_{a,3} - \omega_2\dot{m}_{a,2}\\ &= \dot{m}_a(\omega_3 - \omega_2)\\ &= \left(1.25\ \frac{\text{kg}}{\text{s}}\right)(0.0100 - 0.00109)\left(60\ \frac{\text{s}}{\text{min}}\right)\\ &= 0.668\ \text{kg/min} \quad (0.67\ \text{kg/min})\end{aligned}$$

The answer is (C).

14. The following diagrams apply to this problem.

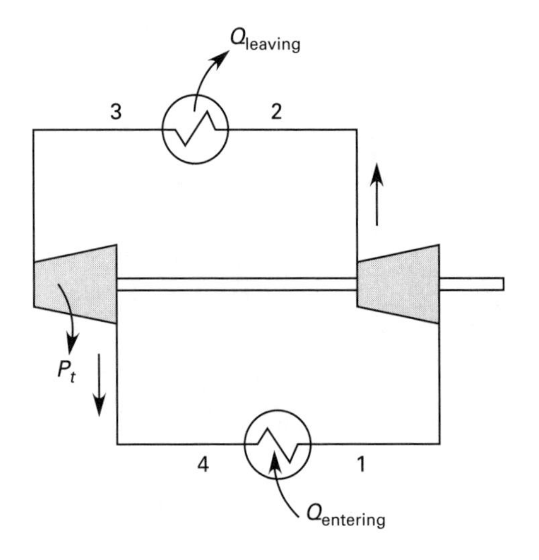

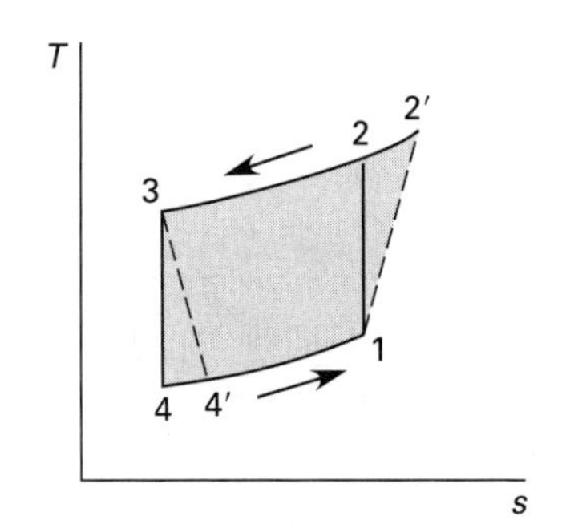

$$\begin{aligned}T_1 &= 25°\text{F} + 460°\\ &= 485°\text{R}\\ T_3 &= 90°\text{F} + 460°\\ &= 550°\text{R}\end{aligned}$$

Find the mass flow rate of air. Since $pV = mRT$,

$$p\dot{V} = \dot{m}RT_1 = \dot{m}\left(\frac{R^*}{\text{MW}}\right)T_1$$

$$\left(14.7\ \frac{\text{lbf}}{\text{in}^2}\right)\left(144\ \frac{\text{in}^2}{\text{ft}^2}\right)\left(2500\ \frac{\text{ft}^3}{\text{min}}\right)\left(60\ \frac{\text{min}}{\text{hr}}\right) = \dot{m}\left(\frac{1545\ \frac{\text{ft-lbf}}{\text{lbmol-}°\text{R}}}{28.97\ \frac{\text{lbm}}{\text{lbmol}}}\right)(485°\text{R})$$

$$\dot{m} = 12{,}276\ \text{lbm/hr}$$

The enthalpy at state 1 can be found using the air properties table at 485°R.

$$h_1 = 115.89\ \text{Btu/lbm}$$

Since the enthalpy at the turbine exit is needed to determine the refrigeration capacity, the turbine power equation can be used to find $h_{4'}$.

$$P_t = \dot{m}(h_3 - h_{4'})$$

This requires that P_t and h_3 be found. Use the air properties table at 550°R to find h_3.

$$h_3 = 131.46\ \text{Btu/lbm}$$

The turbine efficiency is 85%.

$$P_t = 0.85\dot{m}(h_3 - h_4)$$

The enthalpy at state 4 can be found using relative pressures (p_r).

$$\frac{p_3}{p_4} = \frac{p_{r3}}{p_{r4}}$$

The relative pressure at state 3 can be found using the air properties table at 550°R.

$$\begin{aligned}p_{r3} &= 1.4801\\ p_{r4} &= \left(\frac{p_3}{p_4}\right)p_{r3} = \left(\frac{14.7\ \frac{\text{lbf}}{\text{in}^2}}{44.1\ \frac{\text{lbf}}{\text{in}^2}}\right)(1.4801)\\ &= 0.4934\end{aligned}$$

The air properties table shows that p_{r4} lies between 400°R and 420°R. Interpolating to find h_4 gives

$$h_4 = 95.94\ \text{Btu/lbm}$$

The work done by the turbine is

$$P_t = (0.85)\left(12{,}276\ \frac{\text{lbm}}{\text{hr}}\right)\left(131.46\ \frac{\text{Btu}}{\text{lbm}} - 95.94\ \frac{\text{Btu}}{\text{lbm}}\right)$$
$$= 370{,}637\ \text{Btu/hr}$$

Find $h_{4'}$.

$$P_t = \dot{m}(h_3 - h_{4'})$$
$$370{,}637\ \frac{\text{Btu}}{\text{hr}} = \left(12{,}276\ \frac{\text{lbm}}{\text{hr}}\right)\left(131.45\ \frac{\text{Btu}}{\text{lbm}} - h_{4'}\right)$$
$$h_{4'} = 101.26\ \text{Btu/lbm}$$

The refrigeration capacity of the system is

$$\dot{Q}_{\text{ent}} = \dot{m}(h_1 - h_{4'})$$
$$= \left(12{,}276\ \frac{\text{lbm}}{\text{hr}}\right)\left(115.89\ \frac{\text{Btu}}{\text{lbm}} - 101.26\ \frac{\text{Btu}}{\text{lbm}}\right)$$
$$= 179{,}598\ \text{Btu/hr}$$

One ton of refrigeration is defined as the heat required to melt a ton of ice in 24 hours.

$$1\ \text{ton} = 12{,}000\ \text{Btu/hr}$$

$$\dot{Q}_{\text{ent}} = \left(179{,}598\ \frac{\text{Btu}}{\text{hr}}\right)\left(\frac{1\ \text{ton}}{12{,}000\ \frac{\text{Btu}}{\text{hr}}}\right)$$
$$= 14.97\ \text{tons}\quad (15\ \text{tons})$$

The answer is (A).

15. The dimensions of 2½ in schedule-40 steel pipe, and the densities of steel and water are

$$D = 2.875\ \text{in}$$
$$d = 2.469\ \text{in}$$
$$A = 4.788\ \text{in}^2$$
$$\gamma_{\text{steel}} = 490\ \text{lbf/ft}^3\quad (0.28\ \text{lbf/in}^3)$$
$$\gamma_{\text{water}} = 62.4\ \text{lbf/ft}^3\quad (0.0361\ \text{lbf/in}^3)$$

Calculate the weights of pipe and water and the weight, w, per unit length.

$$\frac{w_{\text{pipe}}}{L} = \rho_{\text{steel}}\left(\frac{\pi}{4}(D^2 - d^2)\right)$$
$$= \left(0.28\ \frac{\text{lbf}}{\text{in}^3}\right)\left(\frac{\pi}{4}\left((2.875\ \text{in})^2 - (2.469\ \text{in})^2\right)\right)$$
$$= 0.477\ \text{lbf/in}$$
$$\frac{w_{\text{water}}}{L} = \rho_{\text{water}} A$$
$$= \left(0.0361\ \frac{\text{lbf}}{\text{in}^3}\right)(4.788\ \text{in}^2)$$
$$= 0.173\ \text{lbm/in}$$

$$w = w_{\text{pipe}} + w_{\text{water}}$$
$$= 0.477\ \frac{\text{lbf}}{\text{in}} + 0.173\ \frac{\text{lbf}}{\text{in}}$$
$$= 0.65\ \text{lbf/in}$$

Calculate the moment of inertia for bending.

$$I = \frac{\pi}{64}(D^4 - d^4)$$
$$= \left(\frac{\pi}{64}\right)\left((2.875\ \text{in})^4 - (2.469\ \text{in})^4\right)$$
$$= 1.53\ \text{in}^4$$

Calculate the deflection for a simply supported beam carrying a uniformly distributed load. The modulus of elasticity for steel is $E = 30 \times 10^6\ \text{lbf/in}^2$.

$$y_{\text{max}} = -\frac{5wL^4}{384EI}$$
$$= -\frac{(5)\left(0.65\ \frac{\text{lbf}}{\text{in}}\right)\left((20\ \text{ft})\left(12\ \frac{\text{in}}{\text{ft}}\right)\right)^4}{(384)\left(30 \times 10^6\ \frac{\text{lbf}}{\text{in}^2}\right)(1.53\ \text{in}^4)}$$
$$= 0.611\ \text{in}\quad (0.61\ \text{in})$$

The answer is (D).

16. For a steady, laminar, incompressible flow, the pressure drop is given by

$$\Delta p = \frac{128\mu\dot{V}L}{\pi d^4}$$
$$= \frac{(128)\left(26.37 \times 10^{-6}\ \frac{\text{lbf-sec}}{\text{ft}^3}\right) \times \left(3 \times 10^{-5}\ \frac{\text{ft}^3}{\text{sec}}\right)(3.4\ \text{ft})}{\pi(0.018\ \text{in})^4\left(\frac{1\ \text{ft}}{12\ \text{in}}\right)^4}$$
$$= 21{,}647\ \text{lbf/ft}^2\quad (22{,}000\ \text{lbf/ft}^2)$$

The answer is (C).

17. The chart shows the day-by-day performance of the production crew. This makes it a *run chart* (also known as a *run sequence chart*). Run charts may or may not have warning and control limit lines, as does this one. A run chart used for quality control purposes is called a *Shewhart chart* (or a *control chart* or a *process-behavior chart*). Run charts were introduced as a quality control tool in the 1920s by Walter Shewhart. The chart can help determine whether day-to-day changes in production are normal variations or indications of an underlying problem that should be identified and corrected.

X-bar (moving average), R (range), and p (proportion) charts are also kinds of Shewhart charts, but the data plotted are different from the data in run charts. In a *moving average chart*, each dot represents average performance over a time period. In a *p-chart*, each dot represents the fraction of the day's production that does not conform to a defined standard. In the chart shown, each dot represents one day's production rate, so this is not a moving average chart or a p-chart.

A Gantt chart, which is another project management tool, is a kind of bar chart that shows the start and end dates of elements of a project.

The answer is (D).

18. The dump loader is depreciated each year according to a seven-year MACRS (modified accelerated cost recovery system) schedule. The depreciation fractions for the first three years are

$$D_1 = 14.29\%$$
$$D_2 = 24.29\%$$
$$D_3 = 17.29\%$$

These factors are applied to the purchase price, C, to determine the depreciation in those years. The salvage value is not used in MACRS depreciation calculations.

The before-tax depreciation recovery at three years is

$$\begin{aligned} \mathrm{DR}_{\text{before tax}} &= \\ &= \sum_{t=1}^{3} CD_t(F/P, i, n_t) \\ &= C\sum_{t=1}^{3} D_t(F/P, i, n_t) \\ &= (\$380{,}000)\begin{pmatrix} (0.1429)(F/P, 10\%, 2) \\ +\ (0.2429)(F/P, 10\%, 1) \\ +\ (0.1729)(F/P, 10\%, 0) \end{pmatrix} \end{aligned}$$

The economic factors can be looked up in a factor table.

$$\begin{aligned} \mathrm{DR}_{\text{before tax}} &= (\$380{,}000)\begin{pmatrix} (0.1429)(1.2100) \\ +\ (0.2429)(1.1000) \\ +\ (0.1729)(1.0000) \end{pmatrix} \\ &= \$232{,}940 \end{aligned}$$

This depreciation reduces the company's income taxes by

$$\begin{aligned} \mathrm{DR}_{\text{after tax}} &= d(\mathrm{DR}_{\text{before tax}}) \\ &= (0.36)(\$232{,}940) \\ &= \$83{,}858 \quad (\$84{,}000) \end{aligned}$$

The answer is (C).

19. The moment of inertia is

$$\begin{aligned} J &= \frac{\pi(D^4 - d^4)}{32} = \frac{\pi\left((0.5\ \text{in})^4 - (0.44\ \text{in})^4\right)}{32} \\ &= 0.00246\ \text{in}^4 \end{aligned}$$

The maximum and minimum shear stresses occur at the tube surface. They are

$$\begin{aligned} \tau_{\max} &= \frac{T_{\max} r}{J} = \frac{(30\ \text{in-lbf})(0.25\ \text{in})}{0.00246\ \text{in}^4} \\ &= 3049\ \text{lbf/in}^2 \end{aligned}$$

$$\begin{aligned} \tau_{\min} &= \frac{T_{\min} r}{J} = \frac{(-3\ \text{in-lbf})(0.25\ \text{in})}{0.00246\ \text{in}^4} \\ &= -304.9\ \text{lbf/in}^2 \end{aligned}$$

The alternating and mean shear stresses are

$$\begin{aligned} \tau_{\text{alt}} &= \frac{\tau_{\max} - \tau_{\min}}{2} = \frac{3049\ \frac{\text{lbf}}{\text{in}^2} - \left(-304.9\ \frac{\text{lbf}}{\text{in}^2}\right)}{2} \\ &= 1676\ \text{lbf/in}^2 \end{aligned}$$

$$\begin{aligned} \tau_m &= \frac{\tau_{\max} + \tau_{\min}}{2} = \frac{3049\ \frac{\text{lbf}}{\text{in}^2} + \left(-304.9\ \frac{\text{lbf}}{\text{in}^2}\right)}{2} \\ &= 1372\ \text{lbf/in}^2 \end{aligned}$$

Calculate the safety factor using distortion energy theory and the Goodman fatigue line. Using the distortion energy theory, the endurance strength in shear is

$$\begin{aligned} S_{\text{es}} &= 0.577S_e = (0.577)\left(24{,}000\ \frac{\text{lbf}}{\text{in}^2}\right) \\ &= 13{,}848\ \text{lbf/in}^2 \end{aligned}$$

Similarly, the shear ultimate strength is

$$\begin{aligned} S_{\text{us}} &= 0.577S_u = (0.577)\left(72{,}000\ \frac{\text{lbf}}{\text{in}^2}\right) \\ &= 41{,}544\ \text{lbf/in}^2 \end{aligned}$$

The factor of safety is

$$\begin{aligned} \text{FS} &= \frac{S_{\text{es}}}{\tau_{\text{alt}} + \left(\frac{S_{\text{es}}}{S_{\text{us}}}\right)\tau_m} \\ &= \frac{13{,}848\ \frac{\text{lbf}}{\text{in}^2}}{\left(1676\ \frac{\text{lbf}}{\text{in}^2} + \left(\frac{13{,}848\ \frac{\text{lbf}}{\text{in}^2}}{41{,}544\ \frac{\text{lbf}}{\text{in}^2}}\right) \times \left(1372\ \frac{\text{lbf}}{\text{in}^2}\right)\right)} \\ &= 6.49 \quad (6.5) \end{aligned}$$

The answer is (D).

20. The distance to the neutral axis is

$$c = \frac{t}{2}$$

$$I = \tfrac{1}{12}bt^3 = \left(\frac{1}{12}\right)(0.75 \text{ in})t^3 = 0.0625t^3 \text{ in}$$

The maximum bending stress, $\sigma_{\max}$, is

$$\sigma_{\max} = \frac{Mc}{I} = \frac{(1 \text{ lbf})(4.0 \text{ in})\left(\frac{t}{2}\right)}{0.0625t^3 \text{ in}} = \frac{32 \text{ lbf}}{t^2}$$

For repeated, one-direction stress,

$$\sigma_m = \sigma_a = \frac{\sigma_{\max}}{2} = \frac{32 \text{ lbf}}{2t^2} = \frac{16 \text{ lbf}}{t^2}$$

Solve using the Goodman equation.

$$\text{FS} = \frac{S_e}{\sigma_{\text{alt}} + \left(\frac{S_e}{S_u}\right)\sigma_m}$$

$$3 = \frac{40{,}000 \ \frac{\text{lbf}}{\text{in}^2}}{\frac{16 \text{ lbf}}{t^2} + \left(\frac{40{,}000 \ \frac{\text{lbf}}{\text{in}^2}}{100{,}000 \ \frac{\text{lbf}}{\text{in}^2}}\right)\left(\frac{16 \text{ lbf}}{t^2}\right)} = 1786t^2 \text{ in}^{-2}$$

$$t = 0.041 \text{ in}$$

The answer is (C).

21. Treating the weld as a line, calculate the area, A_w, and section modulus, S_w. (This is equivalent to assuming the weld throat, t_e, is 1.)

$$A_w = \pi D = \pi(3 \text{ in}) = 9.425 \text{ in}$$

$$S_w = \pi r^2 = \frac{\pi D^2}{4} = \frac{\pi(3 \text{ in})^2}{4} = 7.069 \text{ in}^2$$

Calculate the shear and bending loads per unit length of weld.

$$f_{\text{shear}} = \frac{V}{A_w} = \frac{P}{9.425 \text{ in}}$$

$$f_{\text{bending}} = \frac{M}{S_w} = \frac{P(4 \text{ in})}{7.069 \text{ in}^2}$$

The shear load is assumed to be uniform across the weld. The maximum bending load occurs perpendicular to the applied load, parallel to the tube axis, and at the outside diameter. Combine the shear and bending loads to find maximum load per unit leg.

$$f_{\max} = \sqrt{f_{\text{shear}}^2 + f_{\text{bending}}^2} = \sqrt{\left(\frac{P}{9.425 \text{ in}}\right)^2 + \left(\frac{P(4 \text{ in})}{7.069 \text{ in}^2}\right)^2} = \sqrt{\left(0.106P \text{ in}^{-1}\right)^2 + \left(0.566P \text{ in}^{-1}\right)^2} = 0.576P \text{ in}^{-1}$$

Calculate the allowable load.

$$ft_e = \left(10{,}400 \ \frac{\text{lbf}}{\text{in}^2}\right)\left(\frac{3}{16} \text{ in}\right) = 0.576P \text{ in}^{-1}$$

$$P \le 3385 \text{ lbf} \quad (3400 \text{ lbf})$$

The answer is (B).

22. The power required to generate the required amount of cooling can be found using

$$P = \dot{m}(h_2 - h_1)$$

The states are shown below.

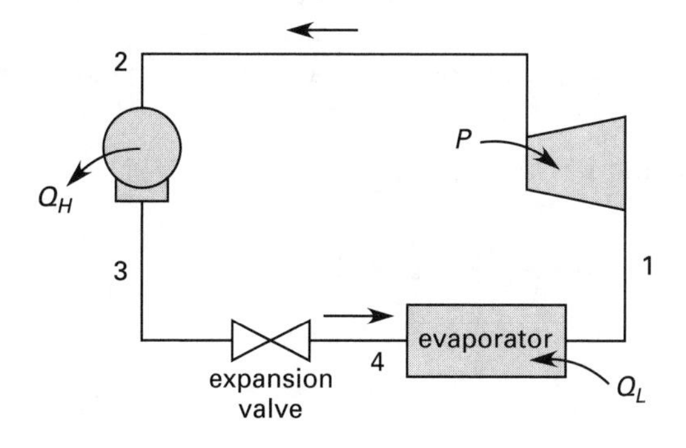

For state 1 (saturated vapor),

$$T_1 = 40°\text{F}$$

Using Freon-12 tables,

$$h_1 = 82.71 \text{ Btu/lbm}$$
$$s_1 = 0.16833 \text{ Btu/lbm-°R}$$

For state 2 (the superheat region), h_2 = the enthalpy of the Freon leaving the compressor.

$$p_2 = p_3$$

For an ideal cycle,

$$s_2 = s_1 = 0.16833 \text{ Btu/lbm-°R}$$

For state 3 (saturated liquid),

$$T_3 = 90\text{°F}$$

Using Freon-12 tables,

$$h_3 = 28.7 \text{ Btu/lbm}$$

The saturation pressure at 90°F is

$$p_3 = 114.3 \text{ psia}$$

Now that p_3 is known, the pressure at state 2 is also known to be 114.3 psia. The enthalpy at state 2 is

$$h_2 = 88.3 \text{ Btu/lbm}$$

The mass flow rate, $\dot{m}$, can be found using

$$\dot{Q}_L = \dot{m}(h_1 - h_4)$$

The throttling valve represents a constant enthalpy process, so

$$h_4 = h_3 = 28.7 \text{ Btu/lbm}$$

$\dot{Q}_L$ is the amount of cooling required. One ton of refrigeration is defined as the heat required to melt a ton of ice in 24 hr. One ton equals 12,000 Btu/hr, so 3 tons equals 36,000 Btu/hr.

Substitute into the equation for $\dot{Q}_L$.

$$36{,}000 \ \frac{\text{Btu}}{\text{hr}} = \dot{m}\left(82.71 \ \frac{\text{Btu}}{\text{lbm}} - 28.7 \ \frac{\text{Btu}}{\text{lbm}}\right)$$
$$\dot{m} = 666.5 \text{ lbm/hr}$$

The power is

$$\begin{aligned} P &= \dot{m}(h_2 - h_1) \\ &= \left(666.5 \ \frac{\text{lbm}}{\text{hr}}\right)\left(88.3 \ \frac{\text{Btu}}{\text{lbm}} - 82.71 \ \frac{\text{Btu}}{\text{lbm}}\right) \\ &\quad \times \left(\frac{1 \text{ hp}}{2545 \ \frac{\text{Btu}}{\text{hr}}}\right) \\ &= 1.46 \text{ hp} \quad (1.5 \text{ hp}) \end{aligned}$$

The answer is (B).

23. The Bernoulli equation can be used to solve this problem.

$$\frac{p_1}{\gamma} + \frac{\text{v}_1^2}{2g} + z_1 + h_p = \frac{p_2}{\gamma} + \frac{\text{v}_2^2}{2g} + z_2 + h_L$$

$$\frac{\left(100 \ \frac{\text{lbf}}{\text{in}^2}\right)\left(144 \ \frac{\text{in}^2}{\text{ft}^2}\right)}{62.4 \ \frac{\text{lbf}}{\text{ft}^3}} + \frac{\text{v}_1^2}{(2)\left(32.2 \ \frac{\text{ft}}{\text{sec}^2}\right)} + 0 \text{ ft} + h_p$$
$$= \frac{\left(500 \ \frac{\text{lbf}}{\text{in}^2}\right)\left(144 \ \frac{\text{in}^2}{\text{ft}^2}\right)}{62.4 \ \frac{\text{lbf}}{\text{ft}^3}} + \frac{\text{v}_2^2}{(2)\left(32.2 \ \frac{\text{ft}}{\text{sec}^2}\right)} + 30 \text{ ft} + 10 \text{ ft}$$

The velocities in both the suction and discharge lines can be found using

$$\dot{V} = \text{v}A$$

The flow rate in each section of the pipe is 7000 gpm. The velocity in the suction line is

$$\left(7000 \ \frac{\text{gal}}{\text{min}}\right)\left(\frac{1 \text{ ft}^3}{7.48 \text{ gal}}\right)\left(\frac{1 \text{ min}}{60 \text{ sec}}\right) = \text{v}_1\left(\frac{\pi}{4}\right)(20 \text{ in})^2\left(\frac{1 \text{ ft}^2}{144 \text{ in}^2}\right)$$
$$\text{v}_1 = 7.15 \text{ ft/sec}$$

The velocity in the discharge line is

$$\left(7000 \ \frac{\text{gal}}{\text{min}}\right)\left(\frac{1 \text{ ft}^3}{7.48 \text{ gal}}\right)\left(\frac{1 \text{ min}}{60 \text{ sec}}\right) = \text{v}_2\left(\frac{\pi}{4}\right)(12 \text{ in})^2\left(\frac{1 \text{ ft}^2}{144 \text{ in}^2}\right)$$
$$\text{v}_2 = 19.9 \text{ ft/sec}$$

Substitute into the Bernoulli equation and solve for the required pump head.

$$\frac{\left(100 \ \frac{\text{lbf}}{\text{in}^2}\right)\left(144 \ \frac{\text{in}^2}{\text{ft}^2}\right)}{62.4 \ \frac{\text{lbf}}{\text{ft}^3}} + \frac{\left(7.15 \ \frac{\text{ft}}{\text{sec}}\right)^2}{(2)\left(32.2 \ \frac{\text{ft}}{\text{sec}^2}\right)} + 0 \text{ ft} + h_p$$
$$= \frac{\left(500 \ \frac{\text{lbf}}{\text{in}^2}\right)\left(144 \ \frac{\text{in}^2}{\text{ft}^2}\right)}{62.4 \ \frac{\text{lbf}}{\text{ft}^3}} + \frac{\left(19.9 \ \frac{\text{ft}}{\text{sec}}\right)^2}{(2)\left(32.2 \ \frac{\text{ft}}{\text{sec}^2}\right)} + 30 \text{ ft} + 10 \text{ ft}$$
$$h_p = 968 \text{ ft}$$

The required power of the pump can be found using

$$\begin{aligned} P &= \dot{V}\gamma h_p \\ &= \left(7000\ \frac{\text{gal}}{\text{min}}\right)\left(\frac{1\ \text{ft}^3}{7.48\ \text{gal}}\right)\left(\frac{1\ \text{min}}{60\ \text{sec}}\right) \\ &\quad \times \left(62.4\ \frac{\text{lbf}}{\text{ft}^3}\right)(968\ \text{ft})\left(\frac{1\ \text{hp}}{550\ \frac{\text{ft-lbf}}{\text{sec}}}\right) \\ &= 1713\ \text{hp} \quad (1700\ \text{hp}) \end{aligned}$$

The answer is (D).

24. The additional power required can be determined using

$$P = F_D\text{v}$$

F_D is the drag force caused by the carrier, and v is the velocity of air relative to the car.

The drag force on the carrier is

$$F_D = \frac{\rho C_D A\text{v}^2}{2g_c}$$

The air density is approximately

$$\rho = 0.075\ \text{lbm/ft}^3$$

The carrier will be modeled as a flat plate. The projected area is

$$\begin{aligned} A &= (35\ \text{in})(18\ \text{in})\left(\frac{1\ \text{ft}^2}{144\ \text{in}^2}\right) \\ &= 4.375\ \text{ft}^2 \end{aligned}$$

The relative wind speed is

$$\begin{aligned} \text{v} &= \left(65\ \frac{\text{mi}}{\text{hr}} + 10\ \frac{\text{mi}}{\text{hr}}\right)\left(5280\ \frac{\text{ft}}{\text{mi}}\right)\left(\frac{1\ \text{hr}}{3600\ \text{sec}}\right) \\ &= 110\ \text{ft/sec} \end{aligned}$$

Determine the drag coefficient, C_D. Compute the Reynolds number.

$$\begin{aligned} \text{Re}_h &= \frac{\text{v}h}{\nu} = \frac{\left(110\ \frac{\text{ft}}{\text{sec}}\right)(18\ \text{in})\left(\frac{1\ \text{ft}}{12\ \text{in}}\right)}{1.58\times 10^{-4}\ \frac{\text{ft}^2}{\text{sec}}} \\ &= 1.04\times 10^6 \end{aligned}$$

To use a drag coefficient chart, the length to height ratio is needed.

$$\frac{L}{h} = \frac{35\ \text{in}}{18\ \text{in}} = 1.94 \approx 2$$

Using the Reynolds number and L/h ratio, the drag coefficient can be found graphically.

$$C_D \approx 1.18$$

The drag force is

$$\begin{aligned} F_D &= \frac{\left(0.075\ \frac{\text{lbm}}{\text{ft}^3}\right)(1.18) \times (4.375\ \text{ft}^2)\left(110\ \frac{\text{ft}}{\text{sec}}\right)^2}{(2)\left(32.2\ \frac{\text{ft-lbm}}{\text{lbf-sec}^2}\right)} \\ &= 72.7\ \text{lbf} \end{aligned}$$

The additional power needed by the car traveling with the carrier is

$$\begin{aligned} P &= F_D\text{v} \\ &= (72.7\ \text{lbf})\left(110\ \frac{\text{ft}}{\text{sec}}\right)\left(\frac{1\ \text{hp}}{550\ \frac{\text{ft-lbf}}{\text{sec}}}\right) \\ &= 14.5\ \text{hp} \quad (15\ \text{hp}) \end{aligned}$$

The answer is (C).

25. The tangential velocity at a radius of 1.5 ft is

$$\text{v}_t = \omega r = \left(188.5\ \frac{\text{rad}}{\text{sec}}\right)(1.5\ \text{ft}) = 282.75\ \text{ft/sec}$$

The air is introduced at three times the tangential velocity, so the velocity of the introduced air is

$$\text{v}_{\text{air}} = 3\text{v}_t = (3)\left(282.75\ \frac{\text{ft}}{\text{sec}}\right) = 848.25\ \text{ft/sec}$$

The velocity of sound is

$$a = \sqrt{kg_cRT}$$

Air temperature is not given, however, so an approximate average value must be assumed. At normal ambient conditions, the speed of sound is about 1125 ft/sec.

The Mach number is

$$\text{M} = \frac{\text{v}}{a} = \frac{848.25\ \frac{\text{ft}}{\text{sec}}}{1125\ \frac{\text{ft}}{\text{sec}}} = 0.754 \quad (0.75)$$

The answer is (D).

26. The total resistance of the window glass includes the glass R-value plus the indoor and outdoor surface resistances.

$$R_{total} = R_{outdoor\ air\ film} + R_{glass} + R_{indoor\ air\ film}$$

The temperature drop between indoor and outdoor temperatures at a point in an assembly is proportional to the R-value at that point relative to the total R-value.

$$\frac{\Delta T_{component}}{\Delta T_{total}} = \frac{R_{component}}{R_{total}}$$
$$\Delta T_{component} = \frac{\Delta T_{total} R_{component}}{R_{total}}$$

The total resistance of the glass is the sum of the resistance of the glass and the reciprocal of the indoor and outdoor surface conductances.

$$\begin{aligned} R_{total} &= \frac{1}{6.0\ \frac{\text{Btu}}{\text{hr-ft}^2\text{-}^\circ\text{F}}} + 1.5\ \frac{\text{hr-ft}^2\text{-}^\circ\text{F}}{\text{Btu}} + \frac{1}{1.46\frac{\text{Btu}}{\text{hr-ft}^2\text{-}^\circ\text{F}}} \\ &= 2.35\ \text{hr-ft}^2\text{-}^\circ\text{F/Btu} \end{aligned}$$

The indoor surface temperature of the glass is calculated by subtracting the temperature drop due to the resistance of the indoor air film from room temperature.

$$\begin{aligned} \Delta T_{indoor\ air\ film} &= (70^\circ\text{F} - (-40^\circ\text{F}))\left(\frac{\dfrac{1}{1.46\ \frac{\text{Btu}}{\text{hr-ft}^2\text{-}^\circ\text{F}}}}{2.35\ \frac{\text{hr-ft}^2\text{-}^\circ\text{F}}{\text{Btu}}}\right) \\ &= 32.06^\circ\text{F} \end{aligned}$$

$$\begin{aligned} T_{surface} &= T_{room} - \Delta T_{indoor\ air\ film} \\ &= 70^\circ\text{F} - 32.06^\circ\text{F} \\ &= 37.94^\circ\text{F} \quad (38^\circ\text{F}) \end{aligned}$$

To avoid condensation, the room air dew point must be less than 38°F. The corresponding relative humidity at room temperature is determined by locating 38°F on the saturation curve of the psychrometric chart and following a line of constant humidity to the right to 70°F.

This state point corresponds to approximately 31% relative humidity.

The answer is (A).

27. The hoop stress, or circumferential stress, is

$$\sigma_h = \frac{pr}{t}$$

The longitudinal stress is half this value.

$$\sigma_l = \frac{pr}{2t}$$

The radius of the tank is

$$\begin{aligned} r &= \frac{d}{2} = \frac{300\ \text{mm}}{2} \\ &= 150\ \text{mm} \end{aligned}$$

The principal stresses are the hoop and longitudinal stresses, and there is no torsion, τ, on the tank, so the stress at an orientation of 55° is

$$\begin{aligned} \sigma_\theta &= \frac{\sigma_h + \sigma_l}{2} + \left(\frac{\sigma_h - \sigma_l}{2}\right)\cos 2\theta + \tau \sin 2\theta \\ &= \frac{\frac{pr}{t} + \frac{pr}{2t}}{2} + \left(\frac{\frac{pr}{t} - \frac{pr}{2t}}{2}\right)\cos 2\theta + \tau \sin 2\theta \\ &= \frac{3pr}{4t} + \left(\frac{pr}{4t}\right)\cos 2\theta + \tau \sin 2\theta \\ \sigma_{55^\circ} &= \frac{3pr}{4t} + \left(\frac{pr}{4t}\right)\cos(2)(55^\circ) + (0\ \text{Pa})\sin(2)(55^\circ) \\ &= 0.665pr/t \end{aligned}$$

The thickness, t, that is needed to limit the stress, σ_{55°, to 115 MPa is

$$\begin{aligned} t &= \frac{0.665pr}{\sigma_{55^\circ}} \\ &= \frac{(0.665)(10\ \text{MPa})(150\ \text{mm})}{115\ \text{MPa}} \\ &= 8.67\ \text{mm} \quad (8.7\ \text{mm}) \end{aligned}$$

The answer is (A).

28. For the book-shelf interface, the coefficient of static friction, μ_s, is

$$\begin{aligned} \mu_s &= \tan\theta = \tan 37^\circ \\ &= 0.753 \quad (0.75) \end{aligned}$$

The answer is (D).

29. $T_1 = 100^\circ\text{F} + 460^\circ = 560^\circ\text{R}$

n is the polytropic exponent. For a polytropic process,

$$\frac{T_2}{T_1} = \left(\frac{p_2}{p_1}\right)^{\frac{n-1}{n}}$$

$$\frac{T_2}{560^\circ\text{R}} = \left(\frac{40\ \frac{\text{lbf}}{\text{in}^2}}{120\ \frac{\text{lbf}}{\text{in}^2}}\right)^{\frac{1.6-1}{1.6}}$$

$$T_2 = 370.9^\circ\text{R}$$

The work done by the system per lbmol is given by

$$\begin{aligned} W &= \frac{R(T_1 - T_2)}{n-1} \\ &= \frac{\left(1545 \frac{\text{ft-lbf}}{\text{lbmol-°R}}\right)(560\text{°R} - 370.9\text{°R})}{1.6 - 1} \\ &= 48.7 \times 10^5 \text{ ft-lbf/lbmol} \end{aligned}$$

The molecular weight of air is 28.97 lbm/lbmol. The work per unit mass is

$$\begin{aligned} W &= \frac{48.7 \times 10^5 \frac{\text{ft-lbf}}{\text{lbmol}}}{28.97 \frac{\text{lbm}}{\text{lbmol}}} \\ &= 1.68 \times 10^4 \text{ ft-lbf/lbm} \end{aligned}$$

The answer is (D).

30. Convert the resistivity to square inches. A circular mil is the area of a circle with a diameter of 0.001 in.

$$\begin{aligned} A_{\text{per cmil}} &= \frac{\pi}{4} d^2 = \left(\frac{\pi}{4}\right)(0.001 \text{ in})^2 \\ &= 7.854 \times 10^{-7} \text{ in}^2 \\ \rho &= \left(10 \frac{\Omega\text{-cmil}}{\text{ft}}\right)\left(7.854 \times 10^{-7} \frac{\text{in}^2}{\text{cmil}}\right) \\ &= 7.854 \times 10^{-6} \ \Omega\text{-in}^2/\text{ft} \end{aligned}$$

The cross-sectional area of the copper tubing is

$$\begin{aligned} A_{\text{tube}} &= \frac{\pi}{4}\left(d_0^2 - d_i^2\right) \\ &= \frac{\pi}{4}\left((3.5 \text{ in})^2 - (3.062 \text{ in})^2\right) \\ &= 2.257 \text{ in}^2 \end{aligned}$$

The maximum resistance that will lead to a drop of no more than 0.6 V is

$$R_{\text{max}} = \frac{V_{\text{max}}}{I} = \frac{0.6 \text{ V}}{35 \text{ A}} = 0.01714 \ \Omega$$

The maximum length of the circuit is

$$\begin{aligned} L_{\text{max}} &= \frac{R_{\text{max}} A_{\text{tube}}}{\rho} \\ &= \frac{(0.01714 \ \Omega)(2.257 \text{ in}^2)}{7.854 \times 10^{-6} \frac{\Omega\text{-in}^2}{\text{ft}}} \\ &= 4926 \text{ ft} \end{aligned}$$

Two wires (pipes) are needed to complete the circuit, so the maximum distance between power supply and pump is

$$d_{\text{max}} = \frac{L_{\text{max}}}{2} = \frac{4926 \text{ ft}}{2} = 2463 \text{ ft} \quad (2500 \text{ ft})$$

The answer is (C).

31. Find the change in temperature.

$$c = \frac{\Delta u}{\Delta T}$$

For water, $c \approx 1.0$ Btu/lbm-°F.

$$\Delta u = \left(1.0 \frac{\text{Btu}}{\text{lbm-°F}}\right) \Delta T$$

Multiply by the mass to find the total change in internal energy.

$$\begin{aligned} \Delta U &= m\Delta u = mc\Delta T \\ &= (60 \text{ lbm})\left(1.0 \frac{\text{Btu}}{\text{lbm-°F}}\right)\Delta T \end{aligned}$$

To find ΔU, use the first law of thermodynamics.

$$Q - W = \Delta U$$

To achieve the maximum possible change in temperature, assume the tank is perfectly insulated so that $Q = 0$. W is negative because work is being done on the system.

$$-W = \Delta U$$

The total work, W, done by the mixing motor is

$$\begin{aligned} W &= Pt \\ &= (2 \text{ hp})\left(550 \frac{\frac{\text{ft-lbf}}{\text{sec}}}{\text{hp}}\right)(15 \text{ min})\left(60 \frac{\text{sec}}{\text{min}}\right) \\ &= 990{,}000 \text{ ft-lbf} \\ \Delta U &= (990{,}000 \text{ ft-lbf})\left(\frac{1 \text{ Btu}}{778 \text{ ft-lbf}}\right) \\ &= 1272 \text{ Btu} \end{aligned}$$

Substitute to find ΔT.

$$\begin{aligned} 1272 \text{ Btu} &= (60 \text{ lbm})\left(1.0 \frac{\text{Btu}}{\text{lbm-°F}}\right)\Delta T \\ \Delta T &= 21.2\text{°F} \quad (21\text{°F}) \end{aligned}$$

The answer is (B).

32. The width of the bearing is given as 1¹/₄ in, which is a fractional measurement (as opposed to a decimal measurement). The tolerance for fractional measurements is given in the illustration as ¹/₈ in. In worst-case manufacturing conditions, the bearing would be ¹/₈ in deficient in width, or 1¹/₄ in − ¹/₈ in = 1¹/₈ in, and the shaft nose would be ¹/₈ in excessive in length.

To keep the shaft nose from protruding even under worst-case conditions, its actual length must be no more than 1¹/₈ in. In the worst case, the shaft nose would be ¹/₈ in longer than dimension A, so dimension A can be no more than 1¹/₈ in − ¹/₈ in = 1 in.

The answer is (B).

33. Draw a diagram and a thermal circuit of the system.

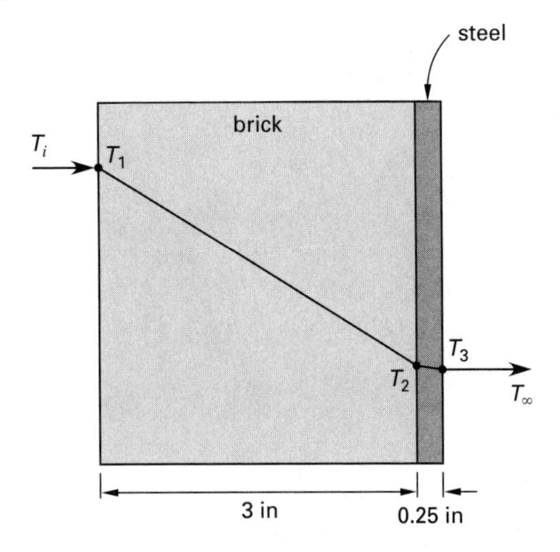

R_{ci} R_{br} R_s R_{co}

T_i — T_1 — T_2 — T_3 — T_∞

Use the given convection coefficient of $\bar{h}_{co} = 1.65$ Btu/hr-ft²-°F. The thermal resistance R_{co} per square foot is

$$R_{co} = \frac{1}{\bar{h}_{co}A} = \frac{1}{\left(1.65\ \frac{\text{Btu}}{\text{hr-ft}^2\text{-°F}}\right)(1\ \text{ft}^2)}$$
$$= 0.606\ \text{hr-°F/Btu}$$
$$k_{br} = 0.58\ \text{Btu-ft/hr-ft}^2\text{-°F}$$
$$k_s = 26\ \text{Btu-ft/hr-ft}^2\text{-°F}$$

(The value of k_s depends on the steel composition, but it does not affect the final result.)

$$R_{br} = \frac{L_{br}}{Ak_{br}} = \frac{(3\ \text{in})\left(\frac{1\ \text{ft}}{12\ \text{in}}\right)}{(1\ \text{ft}^2)\left(0.58\ \frac{\text{Btu-ft}}{\text{hr-ft}^2\text{-°F}}\right)}$$
$$= 0.431\ \text{hr-°F/Btu}$$

$$R_s = \frac{L_s}{Ak_s} = \frac{(0.25\ \text{in})\left(\frac{1\ \text{ft}}{12\ \text{in}}\right)}{(1\ \text{ft}^2)\left(26\ \frac{\text{Btu-ft}}{\text{hr-ft}^2\text{-°F}}\right)}$$
$$= 0.0008\ \text{hr-°F/Btu}$$
$$T_1 = 1000°\text{F}$$
$$T_\infty = 70°\text{F}$$

Since T_1 and T_∞ are the only known temperatures, the heat transfer is

$$q = \frac{T_1 - T_\infty}{R_{br} + R_s + R_{co}}$$
$$= \frac{1000°\text{F} - 70°\text{F}}{0.431\ \frac{\text{hr-°F}}{\text{Btu}} + 0.0008\ \frac{\text{hr-°F}}{\text{Btu}} + 0.606\ \frac{\text{hr-°F}}{\text{Btu}}}$$
$$= 896\ \text{Btu/hr}$$

The outside steel temperature is

$$q = \frac{T_3 - T_\infty}{R_{co}}$$
$$896\ \frac{\text{Btu}}{\text{hr}} = \frac{T_3 - 70°\text{F}}{0.606\ \frac{\text{hr-°F}}{\text{Btu}}}$$
$$T_3 = 613°\text{F} \quad (610°\text{F})$$

The answer is (C).

34. The temperature distribution in a single-pass, parallel-flow heat exchanger is shown.

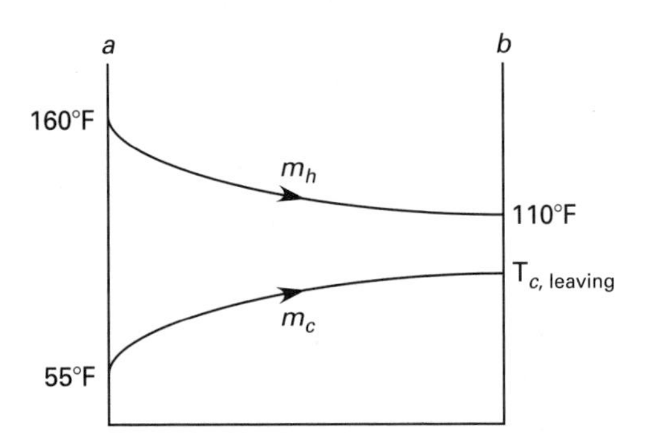

Find $T_{c,\text{leaving}}$.

$$q_h = \dot{m}_h c_{ph} \Delta T_h$$
$$= \left(45{,}000\ \frac{\text{lbm}}{\text{hr}}\right)\left(0.9\ \frac{\text{Btu}}{\text{lbm-°F}}\right)(160°\text{F} - 110°\text{F})$$
$$= 2.025 \times 10^6\ \text{Btu/hr}$$
$$q_c = \dot{m}_c c_{pc} \Delta T_c$$
$$= \left(40{,}000\ \frac{\text{lbm}}{\text{hr}}\right)\left(1\ \frac{\text{Btu}}{\text{lbm-°F}}\right)(T_{c,\text{leaving}} - 55°\text{F})$$

The two heat flow rates are equal, so

$$2.025\times10^6\ \frac{\text{Btu}}{\text{hr}} = \left(40{,}000\ \frac{\text{Btu}}{\text{hr-°F}}\right)(T_{c,\text{leaving}} - 55°\text{F})$$
$$T_{c,\text{leaving}} = 105.6°\text{F}$$

The log mean temperature difference (LMTD) is

$$\begin{aligned}\text{LMTD} &= \frac{\Delta T_a - \Delta T_b}{\ln\dfrac{\Delta T_a}{\Delta T_b}}\\ &= \frac{(160°\text{F} - 55°\text{F}) - (110°\text{F} - 105.6°\text{F})}{\ln\left(\dfrac{160°\text{F} - 55°\text{F}}{110°\text{F} - 105.6°\text{F}}\right)}\\ &= 31.7°\text{F}\end{aligned}$$

The heat transfer in the heat exchanger is

$$q = UA(\text{LMTD})$$
$$2.025\times10^6\ \frac{\text{Btu}}{\text{hr}} = \left(75\ \frac{\text{Btu}}{\text{hr-ft}^2\text{-°F}}\right)A\,(31.7°\text{F})$$
$$A = 851.7\ \text{ft}^2\quad(850\ \text{ft}^2)$$

The answer is (B).

35. Define states 1 and 2′ as shown.

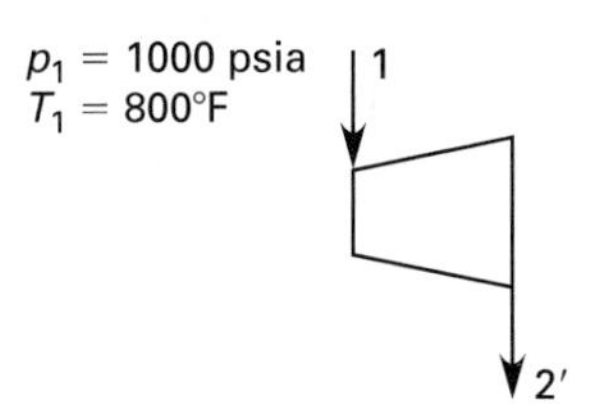

Using steam tables, $h_1 = 1388.5$ Btu/lbm, $s_1 = 1.5664$ Btu/lbm-°R, and $p_2 = 4$ psia. h_2 represents the enthalpy for a turbine that is 100% efficient. Since the turbine is isentropic, $s_1 = s_2$. Using the steam tables, find the appropriate enthalpy and entropy values at state 2′ where $p_2' = 4$ psia.

$$\begin{aligned}h_f &= 120.89\ \text{Btu/lbm}\\ s_f &= 0.21983\ \text{Btu/lbm-°R}\\ h_{fg} &= 1006.4\ \text{Btu/lbm}\\ s_{fg} &= 1.6426\ \text{Btu/lbm-°R}\end{aligned}$$

The steam quality at the turbine exhaust (state 2) for a 100% efficient turbine is found from the entropy relationship.

$$s = s_f + xs_{fg}$$
$$1.5664\ \frac{\text{Btu}}{\text{lbm-°R}} = 0.21983\ \frac{\text{Btu}}{\text{lbm-°R}} + x\left(1.6426\ \frac{\text{Btu}}{\text{lbm-°R}}\right)$$
$$x = 0.82$$

The enthalpy at state 2, h_2, is

$$\begin{aligned}h_2 &= h_f + xh_{fg}\\ &= 120.89\ \frac{\text{Btu}}{\text{lbm}} + (0.82)\left(1006.4\ \frac{\text{Btu}}{\text{lbm}}\right)\\ &= 946.1\ \text{Btu/lbm}\end{aligned}$$

Since the turbine exhaust steam quality is 100%, the enthalpy at state 2′ is equal to the enthalpy of saturated vapor, h_g. From the steam tables at 4 psia,

$$h_{2'} = h_g = 1127.3\ \text{Btu/lbm}$$

The efficiency of the turbine is

$$\begin{aligned}\eta_{\text{turbine}} &= \frac{h_1 - h_{2'}}{h_1 - h_2} = \frac{1388.5\ \dfrac{\text{Btu}}{\text{lbm}} - 1127.3\ \dfrac{\text{Btu}}{\text{lbm}}}{1388.5\ \dfrac{\text{Btu}}{\text{lbm}} - 946.1\ \dfrac{\text{Btu}}{\text{lbm}}}\\ &= 0.59\quad(59\%)\end{aligned}$$

The answer is (A).

Alternate Solution

The enthalpy h_2 can be found from a Mollier diagram, thereby avoiding two calculations. To do this, first find state 1 on the Mollier diagram. It will be at the intersection of the 1000 psia pressure curve and the 800°F curve. Since a 100% efficient turbine is an isentropic process, drop straight down on the diagram until the 4 psia pressure curve is crossed. This intersection represents state 2, and the enthalpy h_2 can be read as approximately 950 Btu/lbm.

36. The cross-sectional area of the cylinder is

$$\begin{aligned}A &= \frac{Q}{\text{v}}\\ &= \frac{\left(3\ \dfrac{\text{gal}}{\text{min}}\right)\left(231\ \dfrac{\text{in}^3}{\text{gal}}\right)}{\left(0.25\ \dfrac{\text{ft}}{\text{sec}}\right)\left(12\ \dfrac{\text{in}}{\text{ft}}\right)\left(60\ \dfrac{\text{sec}}{\text{min}}\right)}\\ &= 3.85\ \text{in}^2\end{aligned}$$

Alternatively, use the standard (and not dimensionally consistent) formula

$$\text{v}_{\text{ft/sec}} = \frac{0.3208Q_{\text{gpm}}}{A_{\text{in}^2}}$$
$$\begin{aligned}A_{\text{in}^2} &= \frac{0.3208Q_{\text{gpm}}}{\text{v}_{\text{ft/sec}}}\\ &= \frac{(0.3208)\left(3\ \dfrac{\text{gal}}{\text{min}}\right)}{0.25\ \dfrac{\text{ft}}{\text{sec}}}\\ &= 3.85\ \text{in}^2\end{aligned}$$

Work is done by the cylinder as it compresses the soap cakes into bars. Work is equal to the force exerted by the cylinder multiplied by the distance it moves.

$$W = Fd$$

The force exerted by the cylinder is

$$F = \eta pA$$

The distance traveled by the cylinder is

$$\begin{aligned} d &= \frac{W}{F} \\ &= \frac{W}{\eta pA} \\ &= \frac{(50 \text{ ft-lbf})\left(12 \ \frac{\text{in}}{\text{ft}}\right)}{(0.85)\left(900 \ \frac{\text{lbf}}{\text{in}^2}\right)(3.85 \text{ in}^2)} \\ &= 0.2037 \text{ in} \quad (0.20 \text{ in}) \end{aligned}$$

The answer is (D).

37. The general combustion reaction for a hydrocarbon, C_nH_m, burning with theoretical air is

$$C_nH_m + xO_2 + 3.76xN_2 \rightarrow aCO_2 + bH_2O + 3.76xN_2$$

$$\begin{aligned} a &= n \quad \text{(from carbon balance)} \\ 2b &= m \quad \text{(from hydrogen balance)} \\ x &= n + \frac{m}{4} \quad \text{(from oxygen balance)} \end{aligned}$$

The composition of methane is CH_4, therefore $n = 1$, and $m = 4$, so $a = 1$, $b = 2$, and $x = 2$. The combustion reaction, then, for combustion of methane in theoretical air is given by

$$\begin{aligned} &CH_4 + 2O_2 + (3.76)(2)N_2 \\ &\quad \rightarrow CO_2 + 2H_2O + (3.76)(2)N_2 \end{aligned}$$

For 125% theoretical air, the reaction is

$$\begin{aligned} &CH_4 + (1.25)(2)O_2 + (1.25)(7.52)N_2 \\ &\quad \rightarrow CO_2 + 2H_2O + (0.25)(2)O_2 + (1.25)(7.52)N_2 \end{aligned}$$

The required amounts of fuel and air are

$$\begin{aligned} m_f &= (1)\left(16 \ \frac{\text{lbm}}{\text{mol}}\right) \\ &= 16 \text{ lbm/mol} \end{aligned}$$

$$\begin{aligned} m_a &= ((1.25)(2) + (1.25)(7.52))\left(28.96 \ \frac{\text{lbm}}{\text{mol}}\right) \\ &= 344.6 \text{ lbm/mol} \end{aligned}$$

The air/fuel ratio is

$$\begin{aligned} R_{\text{af}} &= \frac{m_a}{m_f} = \frac{344.6 \ \frac{\text{lbm}}{\text{mol}}}{16 \ \frac{\text{lbm}}{\text{mol}}} \\ &= 21.5{:}1 \quad (22{:}1) \end{aligned}$$

The answer is (C).

38. Both the hole in the bearing and the shaft nose are dimensioned in decimals. The tolerance for decimal measurements is given in the illustration as 0.005 in. Under worst-case manufacturing conditions, the bearing hole would be 0.005 in deficient in diameter, or

$$b_{\text{bearing hole, worst}} = 0.600 \text{ in} - 0.005 \text{ in} = 0.595 \text{ in}$$

Under worst-case conditions, the shaft nose would be 0.005 in excessive in diameter, or

$$b_{\text{shaft nose, worst}} = 0.590 \text{ in} + 0.005 \text{ in} = 0.595 \text{ in}$$

The minimum clearance, then, is

$$\begin{aligned} C &= b_{\text{bearing hole, worst}} - b_{\text{shaft nose, worst}} \\ &= 0.595 \text{ in} - 0.595 \text{ in} \\ &= 0 \text{ in} \end{aligned}$$

A fit in which the shaft will always fit inside the hole is called a *clearance fit*. There is no interference between the parts, so this is not an *interference fit* (also known as a *press fit*). The term "tolerance fit" has no specific meaning in engineering.

The answer is (A).

39. Determine the fluid velocity through the pipe.

$$\begin{aligned} \text{v} &= \frac{\dot{V}}{A} = \frac{4\dot{V}}{\pi D^2} \\ &= \frac{(4)\left(0.185 \ \frac{\text{m}^3}{\text{min}}\right)}{\pi(0.019 \text{ m})^2\left(60 \ \frac{\text{s}}{\text{min}}\right)} \\ &= 10.9 \text{ m/s} \end{aligned}$$

The total minor loss is the sum of the loss contributions of each of the fittings.

$$\begin{aligned} h_{L,\text{minor}} &= \sum h_{\text{fitting}} \\ &= \sum \frac{\text{v}^2}{2g} k_{\text{fitting}} = \frac{\text{v}^2}{2g}(2k_{90^\circ} + k_{45^\circ}) \\ &= \frac{\left(10.9 \ \frac{\text{m}}{\text{s}}\right)^2}{(2)\left(9.81 \ \frac{\text{m}}{\text{s}^2}\right)}((2)(1.5) + 0.4) \\ &= 20.6 \text{ m} \quad (21 \text{ m}) \end{aligned}$$

The answer is (C).

40. Since the problem involves transient heat flow, determine if the lumped capacitance approximation is valid. The average temperature of the copper is

$$\begin{aligned}\overline{T} &= \frac{T_s + T_l}{2} = \frac{450^\circ\text{C} + 200^\circ\text{C}}{2} \\ &= 325^\circ\text{C} \\ T &= 325^\circ\text{C} + 273^\circ \\ &= 598\text{K}\end{aligned}$$

At this temperature, the sphere's conductivity, specific heat, and density are

$$\begin{aligned}k &= 379\ \text{W/m·K} \\ c_p &= 417\ \text{J/kg·K} \\ \rho &= 8933\ \text{kg/m}^3\end{aligned}$$

Calculate the sphere's characteristic length, L_c.

$$\begin{aligned}L_c &= \frac{V}{A_s} = \frac{\frac{4}{3}\pi R^3}{4\pi R^2} \\ &= \frac{R}{3} \\ &= \frac{100\ \text{mm}}{3} \\ &= 33.3\ \text{mm} \quad (33.3 \times 10^{-3}\ \text{m})\end{aligned}$$

Determine the Biot number, Bi.

$$\begin{aligned}\text{Bi} &= \frac{hL_c}{k} = \frac{\left(880\ \frac{\text{W}}{\text{m}^2\text{·K}}\right)(33.3 \times 10^{-3}\ \text{m})}{379\ \frac{\text{W}}{\text{m·K}}} \\ &= 0.0773\end{aligned}$$

Because the Biot number is less than 0.1, the internal thermal resistance of the sphere is negligible compared to the external thermal resistance in the oil bath. Thus the lumped parameter method can be used.

$$T_t = T_\infty + (T_0 - T_\infty)\, e^{-\text{BiFo}}$$

Solve for the Fourier number, Fo, and substitute its definition.

$$\begin{aligned}\text{Fo} &= \frac{-1}{\text{Bi}} \ln\left(\frac{T_t - T_\infty}{T_0 - T_\infty}\right) \\ \frac{kt}{\rho c_p L_c^2} &= \frac{-1}{\text{Bi}} \ln\left(\frac{T_t - T_\infty}{T_0 - T_\infty}\right)\end{aligned}$$

Solve for t.

$$\begin{aligned}t &= \frac{-\rho c_p L_c^2}{k\text{Bi}} \ln\left(\frac{T_t - T_\infty}{T_0 - T_\infty}\right) \\ &= \frac{-\left(8933\ \frac{\text{kg}}{\text{m}^3}\right)\left(417\ \frac{\text{J}}{\text{kg·K}}\right)(3.33 \times 10^{-2}\ \text{m})^2}{\left(379\ \frac{\text{W}}{\text{m·K}}\right)(7.73 \times 10^{-2})} \\ &\quad \times \ln\left(\frac{200^\circ\text{C} - 75^\circ\text{C}}{450^\circ\text{C} - 75^\circ\text{C}}\right) \\ &= 155\ \text{s} \quad (160\ \text{s})\end{aligned}$$

The answer is (B).

Solutions
HVAC and Refrigeration

41. The useful refrigeration provided by the chiller is determined from the change in enthalpy across the evaporator.

$$\dot{q}_{\text{refrigeration}} = \dot{m}_{\text{refrigerant}} (h_{\text{leaving}} - h_{\text{entering}})$$

The enthalpy of the refrigerant entering the evaporator on the low-pressure side will be equal to that of the refrigerant leaving the condenser as a saturated liquid. From refrigerant property tables for R-134a, the enthalpy of saturated liquid at 100 psia is 37.8 Btu/lbm (38 Btu/lbm).

The enthalpy of the refrigerant entering the compressor is determined by locating the state point for saturated vapor at the low-side pressure, extended horizontally 20°F for superheat. The refrigerant temperature at a saturated vapor pressure of 33 psia is 19.8°F (20°F). Extending a horizontal line from the saturated vapor state point at 33 psia to approximately 40°F (midway between the 20°F and 60°F temperature curves) locates the superheat process line. The extension terminates at 110.2 Btu/lbm (110 Btu/lbm).

Rearranging the refrigeration equation,

$$\begin{aligned}\dot{m}_{\text{refrigerant}} &= \frac{\dot{q}_{\text{refrigeration}}}{h_{\text{leaving}} - h_{\text{entering}}} \\ &= \frac{(255 \text{ tons})\left(12{,}000 \, \frac{\frac{\text{Btu}}{\text{hr}}}{\text{ton}}\right)}{110 \, \frac{\text{Btu}}{\text{lbm}} - 38 \, \frac{\text{Btu}}{\text{lbm}}} \\ &= 42{,}500 \text{ lbm/hr} \quad (43{,}000 \text{ lbm/hr})\end{aligned}$$

The answer is (A).

42. The humidity ratio of the mixed air can be determined graphically on the psychrometric chart using the lever rule or from the ratio of outside air to recirculated quantities.

$$\begin{aligned}\omega_{\text{mixed}} = \omega_{\text{return}} &+ \left(\frac{\dot{V}_{\text{outside}}}{\dot{V}_{\text{return}} + \dot{V}_{\text{outside}}}\right) \\ &\times (\omega_{\text{outside}} - \omega_{\text{return}})\end{aligned}$$

From the psychrometric chart, the humidity ratio for outside air at 90°F db and 75°F wb is 0.0153 lbm moisture/lbm dry air.

The humidity ratio for air returning from the room at 75°F db and 50% relative humidity is 0.0093 lbm moisture/lbm dry air.

$$\begin{aligned}\omega_{\text{mixed}} &= 0.0093 \, \frac{\text{lbm water}}{\text{lbm air}} + \left(\frac{2300 \, \frac{\text{ft}^3}{\text{min}}}{7000 \, \frac{\text{ft}^3}{\text{min}} + 2300 \, \frac{\text{ft}^3}{\text{min}}}\right) \\ &\quad \times \left(0.0153 \, \frac{\text{lbm water}}{\text{lbm air}} - 0.0093 \, \frac{\text{lbm water}}{\text{lbm air}}\right) \\ &= 0.0108 \text{ lbm water/lbm air} \\ &\quad (0.011 \text{ lbm moisture/lbm dry air})\end{aligned}$$

The answer is (B).

43. NPSHA represents the positive head available to introduce liquid into the pump without lowering its pressure below the vapor pressure and causing cavitation. This is determined by subtracting any reductions in head from local atmospheric pressure and subtracting the friction head and vapor pressure head of the water.

$$\begin{aligned}\text{NPSHA} &= h_{\text{atmospheric}} \pm h_{\text{static}} - h_{\text{vapor,pressure}} \\ &\quad - h_{\text{friction}}\end{aligned}$$

Atmospheric pressure is known at sea level. Static and friction head are given, and the vapor pressure of the water at 80°F is determined from the steam tables as 0.50683 lbf/in^2. Use $\gamma = 62.4$ lbf/ft^3 and $h = p/\gamma$.

$$\begin{aligned}\text{NPSHA} &= \left(14.7 \, \frac{\text{lbf}}{\text{in}^2}\right)\left(\frac{144 \, \frac{\text{in}^2}{\text{ft}^2}}{62.4 \, \frac{\text{lbf}}{\text{ft}^3}}\right) \\ &\quad - 17.0 \text{ ft} \\ &\quad - \left(0.51 \, \frac{\text{lbf}}{\text{in}^2}\right)\left(\frac{144 \, \frac{\text{in}^2}{\text{ft}^2}}{62.4 \, \frac{\text{lbf}}{\text{ft}^3}}\right) \\ &\quad - 3.0 \text{ ft} \\ &= 12.7 \text{ ft} \quad (13.0 \text{ ft})\end{aligned}$$

The answer is (A).

44. The affinity laws relate initial and final pump flow and horsepower as

$$\frac{P_{\text{final}}}{P_{\text{initial}}} = \left(\frac{Q_{\text{final}}}{Q_{\text{initial}}}\right)^3$$

The initial horsepower is

$$\begin{aligned} P_{\text{initial}} &= \dot{m}\left(\frac{g}{g_c}\right)h\left(\frac{1}{\eta_{\text{pump}}}\right) \\ &= \dot{V}\rho\left(\frac{g}{g_c}\right)h\left(\frac{1}{\eta_{\text{pump}}}\right) \\ &= \left(250\ \frac{\text{gal}}{\text{min}}\right)\left(8.34\ \frac{\text{lbm}}{\text{gal}}\right)\left(\frac{32.2\ \frac{\text{ft}}{\text{sec}^2}}{32.2\ \frac{\text{ft-lbm}}{\text{lbf-sec}^2}}\right) \\ &\quad \times \left(\frac{1\ \text{hp}}{33{,}000\ \frac{\text{ft-lbf}}{\text{min}}}\right)(75\ \text{ft})\left(\frac{1}{0.65}\right) \\ &= 7.29\ \text{hp} \end{aligned}$$

Rearranging the affinity equation, the final flow achievable when the pump is making full use of the 10 hp motor is

$$\begin{aligned} Q_{\text{final}} &= Q_{\text{initial}}\sqrt[3]{\frac{P_{\text{final}}}{P_{\text{initial}}}} \\ &= 250\ \frac{\text{gal}}{\text{min}}\sqrt[3]{\frac{10\ \text{hp}}{7.29\ \text{hp}}} \\ &= 277.8\ \text{gpm} \quad (280\ \text{gpm}) \end{aligned}$$

The answer is (A).

45. The casino has no walls or windows on the building exterior and the facility runs 24 hr/day. This implies that there will not be significant variation in the cooling load.

The lighting load for the space is

$$\begin{aligned} \dot{q}_{\text{lighting}} &= \left(40{,}000\ \text{ft}^2\right)\left(3.75\ \frac{\text{W}}{\text{ft}^2}\right)\left(3.413\ \frac{\frac{\text{Btu}}{\text{hr}}}{\text{W}}\right) \\ &= 511{,}950\ \text{Btu/hr} \end{aligned}$$

The equipment load for the space is

$$\begin{aligned} \dot{q}_{\text{equipment}} &= (80\ \text{kW})\left(3413\ \frac{\frac{\text{Btu}}{\text{hr}}}{\text{kW}}\right) \\ &= 273{,}040\ \text{Btu/hr} \end{aligned}$$

In the ASHRAE *2001 Handbook of Fundamentals*, active people "walking around" contribute sensible and latent loads of 250 Btu/hr and 200 Btu/hr per person, respectively. The sensible and latent people loads for the space are

$$\begin{aligned} &\dot{q}_{\text{people sensible}} \\ &\quad = \left(40{,}000\ \text{ft}^2\right)\left(\frac{120\ \text{people}}{1000\ \text{ft}^2}\right)\left(250\ \frac{\frac{\text{Btu}}{\text{hr}}}{\text{person}}\right) \\ &\quad = 1{,}200{,}000\ \text{Btu/hr} \\ &\dot{q}_{\text{people latent}} \\ &\quad = \left(40{,}000\ \text{ft}^2\right)\left(\frac{120\ \text{people}}{1000\ \text{ft}^2}\right)\left(200\ \frac{\frac{\text{Btu}}{\text{hr}}}{\text{person}}\right) \\ &\quad = 960{,}000\ \text{Btu/hr} \end{aligned}$$

The sensible load for the space is

$$\begin{aligned} \dot{q}_{\text{sensible}} &= \dot{q}_{\text{lighting}} + \dot{q}_{\text{equipment}} + \dot{q}_{\text{people sensible}} \\ &= 511{,}950\ \frac{\text{Btu}}{\text{hr}} + 273{,}040\ \frac{\text{Btu}}{\text{hr}} \\ &\quad + 1{,}200{,}000\ \frac{\text{Btu}}{\text{hr}} \\ &= 1{,}984{,}990\ \text{Btu/hr} \end{aligned}$$

The latent load for the space is

$$\begin{aligned} \dot{q}_{\text{latent}} &= \dot{q}_{\text{steam table}} + \dot{q}_{\text{people latent}} \\ &= 50{,}000\ \frac{\text{Btu}}{\text{hr}} + 960{,}000\ \frac{\text{Btu}}{\text{hr}} \\ &= 1{,}010{,}000\ \text{Btu/hr} \end{aligned}$$

The total load for the space is

$$\begin{aligned} \dot{q}_{\text{total}} &= \dot{q}_{\text{sensible}} + \dot{q}_{\text{latent}} \\ &= 1{,}984{,}990\ \frac{\text{Btu}}{\text{hr}} + 1{,}010{,}000\ \frac{\text{Btu}}{\text{hr}} \\ &= 2{,}994{,}990\ \text{Btu/hr} \quad (3{,}000{,}000\ \text{Btu/hr}) \end{aligned}$$

The answer is (D).

46. The ADPI method relies on the selection of a diffuser whose throw (distance traveled to the point where velocity is reduced to 50 ft/min) is related to the characteristic room dimension by published ratios, which are empirically derived for 9 ft ceilings. For the case of ceiling diffusers, the maximum ADPI is achieved for all cooling load densities when

$$\frac{t_{50}}{L} = 0.8$$

For the ceiling-mounted diffuser in this problem, the characteristic dimension, L, is the distance from the diffuser to the nearest wall, which is 6.0 ft. The correct throw is

$$\begin{aligned} t_{50} &= (0.8)(6.0 \text{ ft}) \\ &= 4.8 \text{ ft} \quad (5 \text{ ft}) \end{aligned}$$

The answer is (A).

47. From the psychrometric chart, the humidity ratio for air at 70°F and 40% relative humidity is

$$\omega = 0.0062 \text{ lbm water/lbm air}$$

As the air in the car cools, the total mass of air and moisture in the car remains the same, making the process one of constant humidity ratio. Start from the initial conditions and proceed left on the chart along a constant humidity ratio line of 0.0062 lbm/lbm to 50°F. The final relative humidity is 80%.

The answer is (C).

48. Assume the thickness of the tube (and, therefore, conductive heat loss) is negligible. For one-dimensional, steady-state, radial convection, the heat flux into the tube, q_r, is

$$\begin{aligned} q_r &= hA\Delta T \\ &= h(2\pi rL)\Delta T \end{aligned}$$

$$\begin{aligned} \Delta T &= \frac{\frac{q}{L}}{2\pi hr} = \frac{73.0 \frac{\text{W}}{\text{m}}}{(2)(\pi)\left(80 \frac{\text{W}}{\text{m}^2\cdot\text{K}}\right)\left(\frac{0.015 \text{ m}}{2}\right)} \\ &= 19.4°\text{C} \end{aligned}$$

Solve for the temperature of the ambient air, T_a.

$$\begin{aligned} \Delta T &= T_a - T_r \\ T_a &= \Delta T + T_r \\ &= 19.4°\text{C} + (-21°\text{C}) \\ &= -1.6°\text{C} \quad (-2.0°\text{C}) \end{aligned}$$

The answer is (C).

49. According to the Darcy equation, the friction pressure loss in a conduit can be calculated as

$$h_f = f\left(\frac{L}{D_h}\right)\left(\frac{\text{v}^2}{2g}\right)$$

The hydraulic diameter, D_h, is defined as

$$\begin{aligned} D_h &= \frac{4A}{P} = \left(\frac{(4)(18 \text{ in})(24 \text{ in})}{(2)(18 \text{ in}) + (2)(24 \text{ in})}\right)\left(\frac{1 \text{ ft}}{12 \text{ in}}\right) \\ &= 1.714 \text{ ft} \end{aligned}$$

The velocity in the duct is the flow divided by the cross-sectional area.

$$\begin{aligned} \text{v} &= \frac{Q}{A} = \frac{\left(4500 \frac{\text{ft}^3}{\text{min}}\right)\left(\frac{1 \text{ min}}{60 \text{ sec}}\right)}{(18 \text{ in})(24 \text{ in})\left(\frac{1 \text{ ft}^2}{144 \text{ in}^2}\right)} \\ &= 25 \text{ ft/sec} \end{aligned}$$

The Darcy equation will give results in units of feet of the fluid of interest. However, it is traditional to measure fluid head of air in terms of "inches of water," which will require additional unit conversions in terms of length and density. To convert the head from inches of air to inches of water, multiply by the ratio of the densities in these two fluids, $\rho_{\text{air}}/\rho_{\text{water}}$.

$$\begin{aligned} h_f &= (0.016)\frac{(240 \text{ ft})\left(25 \frac{\text{ft}}{\text{sec}}\right)^2}{(2)(1.714 \text{ ft})\left(32.2 \frac{\text{ft}}{\text{sec}^2}\right)} \\ &\quad \times \left(12 \frac{\text{in}}{\text{ft}}\right)\left(\frac{0.075 \frac{\text{lbm air}}{\text{ft}^3}}{62.4 \frac{\text{lbm water}}{\text{ft}^3}}\right) \\ &= 0.31 \text{ in water} \end{aligned}$$

The answer is (B).

50. An energy balance must be established for the crawl space. The heat gain from the heated room above through the floor to the crawl space is equal to heat loss from the crawl space to the outdoor environment. The heat loss to the environment includes heat lost through the crawl space wall and heat lost warming up infiltrating air.

$$\begin{aligned} q_{\text{floor}} &= q_{\text{wall}} + q_{\text{inf}} \\ U_f A_f (T_r - T_c) &= U_w A_w (T_c - T_o) + \dot{m}c_p (T_c - T_o) \end{aligned}$$

The subscripts f, r, c, w, and o respectively represent floor, room, crawl space, wall, and outdoor.

To keep the pipes from freezing, the temperature of the crawl space, T_c, must remain above 32°F.

Solving for the outdoor temperature,

$$\begin{aligned} U_f A_f (T_r - T_c) &= U_w A_w T_c - U_w A_w T_o \\ &\quad + \dot{m}c_p T_c - \dot{m}c_p T_o \\ &= T_c(U_w A_w + \dot{m}c_p) \\ &\quad - T_o(U_w A_w + \dot{m}c_p) \\ T_o(U_w A_w + \dot{m}c_p) &= T_c(U_w A_w + \dot{m}c_p) \\ &\quad - U_f A_f (T_r - T_c) \end{aligned}$$

$$T_o = \frac{T_c(U_wA_w + \dot{m}c_p) - U_fA_f(T_r - T_c)}{U_wA_w + \dot{m}c_p}$$

$$= \frac{(32^\circ\text{F})\left(\begin{array}{l}\left(0.15\ \frac{\text{Btu}}{\text{hr-ft}^2\text{-}^\circ\text{F}}\right)(320\ \text{ft}^2)\\ \quad + \left(900\ \frac{\text{ft}^3}{\text{hr}}\right)\left(0.075\ \frac{\text{lbm}}{\text{ft}^3}\right)\\ \quad \times \left(0.24\ \frac{\text{Btu}}{\text{lbm-}^\circ\text{F}}\right)\end{array}\right) - \left(0.05\ \frac{\text{Btu}}{\text{hr-ft}^2\text{-}^\circ\text{F}}\right)(645\ \text{ft}^2)(72^\circ\text{F} - 32^\circ\text{F})}{\left(\begin{array}{l}\left(0.15\ \frac{\text{Btu}}{\text{hr-ft}^2\text{-}^\circ\text{F}}\right)(320\ \text{ft}^2)\\ \quad + \left(900\ \frac{\text{ft}^3}{\text{hr}}\right)\left(0.075\ \frac{\text{lbm}}{\text{ft}^3}\right)\\ \quad \times \left(0.24\ \frac{\text{Btu}}{\text{lbm-}^\circ\text{F}}\right)\end{array}\right)}$$

$$= 11.9^\circ\text{F} \quad (12^\circ\text{F})$$

The answer is (C).

51. From saturated water tables, at 180°F,

$$h_{fg} = 989.9\ \text{Btu/lbm}$$
$$c_p = 1.00\ \text{Btu/lbm-}^\circ\text{F}$$

The energy needed to evaporate the water is

$$E_{\text{vapor}} = m_{\text{vapor}} h_{fg}$$

An energy balance on the system gives

$$m_{\text{vapor}} h_{fg} = m_{\text{remaining}} c_p \Delta T_{\text{remaining}}$$
$$\Delta T_{\text{remaining}} = \frac{m_{\text{vapor}} h_{fg}}{m_{\text{remaining}} c_p}$$

8% of the water evaporates, so

$$\frac{m_{\text{vapor}}}{m_{\text{remaining}}} = \frac{8\%}{92\%} = 0.0870$$

The temperature change in the remaining water is thus

$$\Delta T_{\text{remaining}} = \left(\frac{m_{\text{vapor}}}{m_{\text{remaining}}}\right)\left(\frac{h_{fg}}{c_p}\right) = (0.0870)\left(\frac{989.9\ \frac{\text{Btu}}{\text{lbm}}}{1.00\ \frac{\text{Btu}}{\text{lbm-}^\circ\text{F}}}\right) = 86.1^\circ\text{F}$$

The remaining water has a temperature of

$$T_{\text{remaining}} = T_{\text{initial}} - \Delta T_{\text{remaining}} = 180^\circ\text{F} - 86.1^\circ\text{F} = 93.9^\circ\text{F} \quad (94^\circ\text{F})$$

The answer is (C).

52. The brake horsepower required by the pump is

$$P_{\text{pump}} = \dot{m}\left(\frac{g}{g_c}\right) h \left(\frac{1}{\eta_{\text{pump}}}\right) = \dot{V}\rho\left(\frac{g}{g_c}\right) h \left(\frac{1}{\eta_{\text{pump}}}\right)$$

$$= \left(6.0\ \frac{\text{gal}}{\text{min}}\right)\left(8.34\ \frac{\text{lbm}}{\text{gal}}\right)\left(\frac{32.2\ \frac{\text{ft}}{\text{sec}^2}}{32.2\ \frac{\text{ft-lbm}}{\text{lbf-sec}^2}}\right) \times (140\ \text{ft})\left(\frac{1\ \text{hp}}{33{,}000\ \frac{\text{ft-lbf}}{\text{min}}}\right)\left(\frac{1}{0.60}\right)$$

$$= 0.354\ \text{hp}$$

The solar array area is determined by dividing the power requirement by the product of the incident power available, I, and the array collection efficiency, η.

$$A = \frac{P_{\text{pump}}}{\eta_{\text{array}} I_{\text{incident}}} = \frac{(0.354\ \text{hp})\left(746\ \frac{\text{W}}{\text{hp}}\right)}{(0.12)\left(28\ \frac{\text{W}}{\text{ft}^2}\right)} = 78.6\ \text{ft}^2 \quad (79\ \text{ft}^2)$$

The answer is (D).

53. The saturation efficiency for an air washer describes the completeness to which the dry-bulb temperature is reduced to the theoretical minimum wet-bulb temperature.

$$\eta_{\text{sat}} = \frac{T_{\text{db,air,entering}} - T_{\text{db,air,leaving}}}{T_{\text{db,air,entering}} - T_{\text{wb}}}$$

The leaving dry-bulb temperature for this process is

$$T_{\text{db,air,leaving}} = T_{\text{db,air,entering}} - \eta_{\text{sat}}(T_{\text{db,air,entering}} - T_{\text{wb}}) = 92^\circ\text{F} - (0.84)(92^\circ\text{F} - 57^\circ\text{F}) = 62.6^\circ\text{F}$$

The sensible pre-cooling benefit is

$$\begin{aligned}\dot{q}_{\text{sensible}} &= \dot{m}c_p\left(T_{\text{db,air,entering}} - T_{\text{db,air,leaving}}\right)\\ &= \dot{V}\rho c_p\left(T_{\text{db,air,entering}} - T_{\text{db,air,leaving}}\right)\\ &= \left(20{,}000\ \frac{\text{ft}^3}{\text{min}}\right)\left(60\ \frac{\text{min}}{\text{hr}}\right)\left(0.075\ \frac{\text{lbm}}{\text{ft}^3}\right)\\ &\quad\times\left(0.24\ \frac{\text{Btu}}{\text{lbm-}^\circ\text{F}}\right)(92.0^\circ\text{F} - 62.6^\circ\text{F})\\ &= 635{,}040\ \text{Btu/hr}\end{aligned}$$

The chiller loads are traditionally stated in units of tons of refrigeration. be

$$\begin{aligned}\dot{q}_{\text{sensible}} &= \left(635{,}040\ \frac{\text{Btu}}{\text{hr}}\right)\left(\frac{1\ \text{ton}}{12{,}000\ \frac{\text{Btu}}{\text{hr}}}\right)\\ &= 52.92\ \text{tons}\quad(53\ \text{tons})\end{aligned}$$

The answer is (C).

54. The only load is the heat loss through the brick wall. The heat loss can be calculated as

$$q_{\text{wall}} = U_{\text{wall}}A_{\text{wall}}\Delta T$$

The overall heat transfer coefficient (U-factor) is calculated by taking the reciprocal of the total thermal resistance of the wall assembly, including air surfaces. If the construction of the wall is thermally homogenous (continuous without conductive penetrations), resistance of the wall is the simple sum of the individual resistances. Individual resistances can be obtained from a wide variety of sources, the most comprehensive being the ASHRAE *Handbook of Fundamentals*. Typical resistance values for the individual wall components are listed in the following table.

wall components	resistance $\left(\frac{\text{hr-ft}^2\text{-}^\circ\text{F}}{\text{Btu}}\right)$
8 in brick	0.80
2 in polystyrene	8.00
5/8 in gypsum board	0.56
outdoor air film	0.17
indoor air film (vertical)	0.68
total resistance	10.21

The wall overall heat transfer coefficient, U_{wall}, is

$$\begin{aligned}U_{\text{wall}} &= \frac{1}{R_{\text{wall}}} = \frac{1}{10.21\ \frac{\text{hr-ft}^2\text{-}^\circ\text{F}}{\text{Btu}}}\\ &= 0.098\ \text{Btu/hr-ft}^2\text{-}^\circ\text{F}\end{aligned}$$

The heat loss through the wall is

$$\begin{aligned}q_{\text{wall}} &= U_{\text{wall}}A_{\text{wall}}\Delta T\\ &= \left(0.098\ \frac{\text{Btu}}{\text{hr-ft}^2\text{-}^\circ\text{F}}\right)(800\ \text{ft}^2)(74^\circ\text{F} - 10^\circ\text{F})\\ &= 5018\ \text{Btu/hr}\end{aligned}$$

The heating capacity is

$$\begin{aligned}q_{\text{heater}} &= \left(5018\ \frac{\text{Btu}}{\text{hr}}\right)\left(\frac{1\ \text{kW}}{3412\ \frac{\text{Btu}}{\text{hr}}}\right)(1.25)\\ &= 1.84\ \text{kW}\quad(1.9\ \text{kW})\end{aligned}$$

The answer is (B).

55. The sensible capacity of the coil can be calculated as

$$\begin{aligned}\dot{q}_{\text{sensible}} &= \dot{m}c_p\left(T_{\text{entering,db}} - T_{\text{leaving,db}}\right)\\ &= \dot{V}\rho c_p\left(T_{\text{db,air,entering}} - T_{\text{db,air,leaving}}\right)\end{aligned}$$

The airflow through the coil is the product of the velocity of the air and the cross-sectional area of the coil.

$$\begin{aligned}\dot{V} &= \text{v}A\\ &= \left(450\ \frac{\text{ft}}{\text{min}}\right)(3.0\ \text{ft})(4.0\ \text{ft})\\ &= 5400\ \text{ft}^3/\text{min}\end{aligned}$$

The sensible capacity of the coil is

$$\begin{aligned}\dot{q}_{\text{sensible}} &= \left(5400\ \frac{\text{ft}^3}{\text{min}}\right)\left(60\ \frac{\text{min}}{\text{hr}}\right)\left(0.075\ \frac{\text{lbm}}{\text{ft}^3}\right)\\ &\quad\times\left(0.24\ \frac{\text{Btu}}{\text{lbm-}^\circ\text{F}}\right)(80^\circ\text{F} - 56^\circ\text{F})\\ &= 139{,}968\ \text{Btu/hr}\end{aligned}$$

The grand sensible heat ratio of the coil is the ratio of the sensible capacity to the total capacity.

$$\text{GSHR} = \frac{\dot{q}_{\text{sensible}}}{\dot{q}_{\text{total}}}$$

The total coil capacity is determined from the definition

$$\begin{aligned}\dot{q}_{\text{total}} &= \frac{\dot{q}_{\text{sensible}}}{\text{GSHR}} = \frac{139{,}968\ \frac{\text{Btu}}{\text{hr}}}{0.70}\\ &= 199{,}954\ \text{Btu/hr}\quad(200{,}000\ \text{Btu/hr})\end{aligned}$$

The answer is (D).

56. The COP is the rate of useful cooling divided by the power the compressor requires to accomplish the cooling.

$$\text{COP} = \frac{\dot{q}_{\text{cooling}}}{P}$$

The load rejected to the cooling tower includes both the heat rejected as cooling load and the required compressor energy. The total load rejected to the cooling tower is

$$\begin{aligned}\dot{q}_{\text{condenser}} &= \dot{m}c_p\Delta T_{\text{condenser}} \\ &= \left(2530\ \frac{\text{gal}}{\text{min}}\right)\left(8.34\ \frac{\text{lbm}}{\text{gal}}\right)\left(60\ \frac{\text{min}}{\text{hr}}\right) \\ &\quad \times \left(1.0\ \frac{\text{Btu}}{\text{lbm-°F}}\right)(92\text{°F} - 83\text{°F}) \\ &\quad \times \left(\frac{1\ \text{ton}}{12{,}000\ \frac{\text{Btu}}{\text{hr}}}\right) \\ &= 949.5\ \text{tons} \quad (950\ \text{tons})\end{aligned}$$

The compressor heat is the difference between the total load rejected to the condenser and the cooling capacity of the chiller.

$$\begin{aligned}\dot{q}_{\text{compressor}} &= \dot{q}_{\text{rejected}} - \dot{q}_{\text{chiller}} \\ &= 950\ \text{tons} - 840\ \text{tons} \\ &= 110\ \text{tons}\end{aligned}$$

The chiller COP is

$$\begin{aligned}\text{COP} &= \frac{840\ \text{tons}}{110\ \text{tons}} \\ &= 7.67 \quad (7.7)\end{aligned}$$

The answer is (D).

57. Airfoil fans have the highest efficiency, due to the airfoil contour of each of their blades. The fans with the next-highest efficiency are the backward-inclined and backward-curved fans. Forward-curved fans have the lowest efficiency.

The answer is (D).

58. Convert temperature from Fahrenheit to Rankine, and calculate the initial pressure in the bottle. Obtain the gas constant, R, from a table.

$$\begin{aligned}T_R &= T_{\text{°F}} + 460° = 100\text{°F} + 460° \\ &= 560\text{°R}\end{aligned}$$

$$R_{\text{nitrogen}} = 55.16\ \text{ft-lbf/lbm-°R}$$

$$p = \frac{m_1RT}{V}$$

Calculate the mass remaining in the bottle.

$$\begin{aligned}m_2 &= \frac{pV}{RT} = \frac{\left(150\ \frac{\text{lbf}}{\text{in}^2}\right)\left(12\ \frac{\text{in}}{\text{ft}}\right)^2(6\ \text{ft}^3)}{\left(55.16\ \frac{\text{ft-lbf}}{\text{lbm-°R}}\right)(560\text{°R})} \\ &= 4.19\ \text{lbm}\end{aligned}$$

Calculate the mass released.

$$\begin{aligned}m_{\text{released}} &= m_1 - m_2 \\ &= 5\ \text{lbm} - 4.19\ \text{lbm} \\ &= 0.81\ \text{lbm} \quad (0.8\ \text{lbm})\end{aligned}$$

The answer is (B).

59. The ventilation load is the rate at which sensible and latent heat must be removed from the ventilation air to reduce it to the delivery state.

$$\begin{aligned}\dot{q}_{\text{total}} &= \dot{m}\Delta h \\ &= \dot{V}\rho\left(h_{\text{outdoor}} - h_{\text{delivery}}\right)\end{aligned}$$

From the psychrometric chart at 94°F db and 72°F wb,

$$h_{\text{outdoor}} = 35.6\ \text{Btu/lbm}$$

At 55°F and 30% relative humidity,

$$\begin{aligned}h_{\text{room}} &= 16.0\ \text{Btu/lbm} \\ \dot{q}_{\text{total}} &= \left(850\ \frac{\text{ft}^3}{\text{min}}\right)\left(60\ \frac{\text{min}}{\text{hr}}\right)\left(0.075\ \frac{\text{lbm}}{\text{ft}^3}\right) \\ &\quad \times \left(35.6\ \frac{\text{Btu}}{\text{lbm}} - 16.0\ \frac{\text{Btu}}{\text{lbm}}\right) \\ &= 74{,}970\ \text{Btu/hr} \quad (75{,}000\ \text{Btu/hr})\end{aligned}$$

The answer is (C).

60. The reheat coil must have sufficient heating capacity to offset winter space heat loss, as well as the colder supply air.

$$\dot{q}_{\text{coil}} = \dot{q}_{\text{space}} + \dot{q}_{\text{reheat}}$$

The reheat load is given by

$$\begin{aligned}\dot{q}_{\text{reheat}} &= \dot{m}c_p\left(T_{\text{room}} - T_{\text{supply}}\right) \\ &= \dot{V}_{\text{min}}\rho_{\text{air}}c_p\left(T_{\text{room}} - T_{\text{supply}}\right)\end{aligned}$$

The lowest airflow is the product of the minimum stop fraction and the maximum flow.

$$\begin{aligned}\dot{V}_{\text{min}} &= F_{\text{min}}\dot{V}_{\text{max}} = (0.30)\left(2400\ \frac{\text{ft}^3}{\text{min}}\right) \\ &= 720\ \text{ft}^3\text{/min}\end{aligned}$$

The reset chart indicates that the supply air temperature will be reset to 64°F when the outdoor temperature is 10°F.

$$\begin{aligned}\dot{q}_{\text{reheat}} &= \left(720\ \frac{\text{ft}^3}{\text{min}}\right)\left(60\ \frac{\text{min}}{\text{hr}}\right)\left(0.075\ \frac{\text{lbm}}{\text{ft}^3}\right) \\ &\quad \times \left(0.24\ \frac{\text{Btu}}{\text{lbm-°F}}\right)(72\text{°F} - 64\text{°F}) \\ &= 6220\ \text{Btu/hr}\end{aligned}$$

The total heating coil load is

$$\begin{aligned}\dot{q}_{\text{coil}} &= \dot{q}_{\text{space}} + \dot{q}_{\text{reheat}} \\ &= 45{,}000\ \frac{\text{Btu}}{\text{hr}} + 6220\ \frac{\text{Btu}}{\text{hr}} \\ &= 51{,}220\ \text{Btu/hr} \quad (51{,}000\ \text{Btu/hr})\end{aligned}$$

The answer is (C).

61. The time required for the steam to cool to the ambient temperature is determined by dividing the energy reduction (from 212°F saturated steam to 60°F water) by the rate of heat loss from the pipe.

$$t = \frac{E_{\text{initial}} - E_{\text{final}}}{\dot{q}_{\text{pipe}}}$$

From standard pipe dimension schedules, the interior diameter of a 2 in schedule-40 pipe is 2.037 in. The volume of the steam in the pipe is

$$\begin{aligned}V &= A_i L = \frac{\pi d^2}{4} L \\ &= \left(\frac{\pi}{4}\right)\left(\frac{2.067\ \text{in}}{12\ \frac{\text{in}}{\text{ft}}}\right)^2 (80\ \text{ft}) \\ &= 1.864\ \text{ft}^3\end{aligned}$$

From the saturated steam tables, the specific volume of saturated 212°F steam is

$$\upsilon = 26.80\ \text{ft}^3/\text{lbm}$$

The mass of the steam in the pipe is

$$\begin{aligned}m_{\text{steam}} &= \rho V = \frac{V}{\upsilon} \\ &= \frac{1.864\ \text{ft}^3}{26.80\ \frac{\text{ft}^3}{\text{lbm}}} \\ &= 0.0696\ \text{lbm}\end{aligned}$$

The initial enthalpy of the saturated 212°F steam is 1150.5 Btu/lbm.

Since the steam is initially saturated, any cooling will result in some condensation. At 60°F, all the steam will have condensed, and there will be 0.0696 lbm of 60°F liquid water in the pipe. The enthalpy of 60°F water is approximately 28.08 Btu/lbm. The total energy loss is

$$\begin{aligned}\Delta E &= E_{\text{initial}} - E_{\text{final}} = m\Delta h \\ &= (0.0696\ \text{lbm})\left(1150.5\ \frac{\text{Btu}}{\text{lbm}} - 28.08\ \frac{\text{Btu}}{\text{lbm}}\right) \\ &= 78.12\ \text{Btu}\end{aligned}$$

The time to cool is

$$\begin{aligned}t &= \frac{\Delta E_{\text{steam}}}{\dot{q}_{\text{pipe}}} = \frac{78.12\ \text{Btu}}{\left(6200\ \frac{\text{Btu}}{\text{hr}}\right)\left(\frac{1\ \text{hr}}{60\ \text{min}}\right)} \\ &= 0.756\ \text{min} \quad (1\ \text{min})\end{aligned}$$

The answer is (B).

62. Table 2 in ASHRAE Standard 62 requires 15 ft^3/min of outdoor air per person. The maximum estimated occupancy, according to the same table, is 150 people per 1000 ft^2.

$$\begin{aligned}\dot{V}_{\text{ventilation}} &= \left(\frac{150\ \text{people}}{1000\ \text{ft}^2}\right)(5000\ \text{ft}^2) \\ &\quad \times \left(15\ \frac{\text{ft}^3}{\text{min-person}}\right) \\ &= 11{,}250\ \text{ft}^3/\text{min} \quad (11{,}000\ \text{ft}^3/\text{min})\end{aligned}$$

The answer is (C).

63. The entropy of the steam entering the turbine at 1200°F and 700 psia is

$$s_3 = 1.7686\ \text{Btu/lbm-°R}$$

For maximum efficiency, the entropy of the steam entering the condenser, s_4, has the same value. The quality can be found from the equation

$$\begin{aligned}s_3 &= s_4 \\ &= s_f + x(s_g - s_f) \\ x &= \frac{s_3 - s_f}{s_g - s_f}\end{aligned}$$

s_f and s_g are the entropies of the saturated liquid and vapor, respectively, from a saturated steam table for a pressure of 2 psi.

$$\begin{aligned}x &= \frac{1.7686\ \frac{\text{Btu}}{\text{lbm-°R}} - 0.1750\ \frac{\text{Btu}}{\text{lbm-°R}}}{1.9195\ \frac{\text{Btu}}{\text{lbm-°R}} - 0.1750\ \frac{\text{Btu}}{\text{lbm-°R}}} \\ &= 0.9135\end{aligned}$$

The enthalpy of the steam entering the condenser is

$$h_4 = h_f + x(h_g - h_f)$$

h_f and h_g are the enthalpies of the saturated liquid and vapor, respectively, and are taken from a saturated steam table for a pressure of 2 psia.

$$\begin{aligned} h_4 &= 94.02\ \frac{\text{Btu}}{\text{lbm}} + (0.9135)\left(1115.8\ \frac{\text{Btu}}{\text{lbm}} - 94.02\ \frac{\text{Btu}}{\text{lbm}}\right) \\ &= 1027.4\ \text{Btu/lbm} \end{aligned}$$

The enthalpy between the turbine and boiler can be found in a superheated steam table for 1200°F and 700 psia, and is

$$h_3 = 1625.9\ \text{Btu/lbm}$$

The enthalpy of the saturated 2 psia liquid is

$$h_1 = h_f = 94.02\ \text{Btu/lbm}$$

The efficiency of the cycle is

$$\begin{aligned} \eta &= \frac{W}{Q} = \frac{h_3 - h_4}{h_3 - h_1} \\ &= \frac{1625.9\ \frac{\text{Btu}}{\text{lbm}} - 1027.4\ \frac{\text{Btu}}{\text{lbm}}}{1625.9\ \frac{\text{Btu}}{\text{lbm}} - 94.02\ \frac{\text{Btu}}{\text{lbm}}} \\ &= 0.391 \quad (39\%) \end{aligned}$$

The answer is (C).

64. Since the temperatures of water entering and leaving the condenser are fixed, the required flow is determined by the total load to be rejected, which includes the load absorbed by the evaporator in addition to the heat added by the compressor. The relationship for the condenser water temperature rise is

$$\dot{q}_{\text{rejected}} = \dot{m}_{\text{condenser water}} c_p \Delta T$$

The ΔT term represents the water temperature change, and the load to be rejected is

$$\dot{q}_{\text{rejected}} = \dot{q}_{\text{refrigeration}} + \dot{q}_{\text{comp heat}}$$

The required condenser water can be determined as

$$\dot{m}_{\text{condenser water}} = \frac{\dot{q}_{\text{refrigeration}} + \dot{q}_{\text{comp heat}}}{c_p \Delta T}$$

The refrigeration capacity of the chiller is the refrigerant mass flow rate times the refrigeration effect.

$$\begin{aligned} \dot{q}_{\text{refrigeration}} &= \dot{m}_{\text{refrigerant}}(h_{\text{evap entering}} - h_{\text{evap leaving}}) \\ &= \dot{m}_{\text{refrigerant}}(h_{\text{evap entering}} - h_{\text{comp entering}}) \end{aligned}$$

The enthalpy of the refrigerant mixture entering the chiller evaporator will be equal to that of the saturated liquid leaving the chiller condenser, since there is no change in enthalpy when the refrigerant goes through the throttling valve. Refrigerant enthalpies are found in an R-22 property table. The enthalpy of the saturated liquid at 90°F is

$$h_{\text{evap entering}} = 36.12\ \text{Btu/lbm}$$

The enthalpy of the refrigerant entering the compressor is equal to the enthalpy of the saturated vapor at 10°F, since there is no superheat. The enthalpy of the saturated vapor at 10°F is

$$h_{\text{comp entering}} = 105.27\ \text{Btu/lbm}$$

The useful refrigeration is

$$\begin{aligned} \dot{q}_{\text{refrigeration}} &= \left(117{,}000\ \frac{\text{lbm}}{\text{hr}}\right) \\ &\quad \times \left(105.27\ \frac{\text{Btu}}{\text{lbm}} - 36.121\ \frac{\text{Btu}}{\text{lbm}}\right) \\ &= 8{,}090{,}433\ \text{Btu/hr} \end{aligned}$$

Based on the definition of COP, the compressor work can be determined.

$$\begin{aligned} \dot{q}_{\text{comp}} &= \frac{\dot{q}_{\text{refrigeration}}}{\text{COP}} = \frac{8{,}090{,}433\ \frac{\text{Btu}}{\text{hr}}}{5.5} \\ &= 1{,}470{,}988\ \text{Btu/hr} \end{aligned}$$

The required rate of heat removal is

$$\begin{aligned} \dot{q}_{\text{rejected}} &= \dot{q}_{\text{refrigeration}} + \dot{q}_{\text{comp}} \\ &= 8{,}090{,}433\ \frac{\text{Btu}}{\text{hr}} + 1{,}470{,}988\ \frac{\text{Btu}}{\text{hr}} \\ &= 9{,}561{,}421\ \text{Btu/hr} \end{aligned}$$

The condenser water flow that is required to remove the heat is

$$\dot{m}_{\text{condenser}} = \rho \dot{V}_{\text{condenser}} = \frac{\dot{q}_{\text{rejected}}}{c_p \Delta T}$$

$$\begin{aligned} \dot{V}_{\text{condenser}} &= \frac{\dot{q}_{\text{rejected}}}{\rho c_p \Delta T} \\ &= \frac{\left(9{,}561{,}421\ \frac{\text{Btu}}{\text{hr}}\right)\left(\frac{1\ \text{hr}}{60\ \text{min}}\right)}{\left(8.34\ \frac{\text{lbm}}{\text{gal}}\right)\left(1.0\ \frac{\text{Btu}}{\text{lbm-°F}}\right)(95\text{°F} - 85\text{°F})} \\ &= 1910.8\ \text{gpm} \quad (1900\ \text{gpm}) \end{aligned}$$

The answer is (C).

65. ASHRAE recommends in its *Systems and Equipment Handbook* that the required volume for closed tanks with air-water interfaces should be

$$V_{\text{tank}} = V_{\text{system}} \left(\frac{\left(\frac{v_2}{v_1} - 1\right) - 3\alpha\Delta T}{\frac{p_{\text{atm}}}{p_1} - \frac{p_{\text{atm}}}{p_2}} \right)$$

The specific volume of water at 50°F is

$$v_1 = 0.01602 \text{ ft}^3/\text{lbm}$$

The specific volume of water at 105°F is

$$v_2 = 0.01615 \text{ ft}^3/\text{lbm}$$

The minimum absolute pressure, p_1, for the tank is

$$\begin{aligned} p_1 &= p_{\text{gage}} + p_{\text{atm}} \\ &= 10 \, \frac{\text{lbf}}{\text{in}^2} + 14.7 \, \frac{\text{lbf}}{\text{in}^2} \\ &= 24.7 \text{ lbf/in}^2 \end{aligned}$$

The maximum absolute pressure, p_2, for the tank is

$$\begin{aligned} p_2 &= 23 \, \frac{\text{lbf}}{\text{in}^2} + 14.7 \, \frac{\text{lbf}}{\text{in}^2} \\ &= 37.7 \text{ lbf/in}^2 \end{aligned}$$

The expansion tank size is

$$\begin{aligned} V_{\text{tank}} &= (1500 \text{ gal}) \left(\frac{\left(\frac{0.01615 \, \frac{\text{ft}^3}{\text{lbm}}}{0.01602 \, \frac{\text{ft}^3}{\text{lbm}}} - 1 \right) - (3)\left(0.0000065 \, \frac{\text{in}}{\text{in-°F}}\right) \times (105°\text{F} - 50°\text{F})}{\frac{14.7 \, \frac{\text{lbf}}{\text{in}^2}}{24.7 \, \frac{\text{lbf}}{\text{in}^2}} - \frac{14.7 \, \frac{\text{lbf}}{\text{in}^2}}{37.7 \, \frac{\text{lbf}}{\text{in}^2}}} \right) \\ &= 51.47 \text{ gal} \quad (50 \text{ gal}) \end{aligned}$$

The answer is (C).

66. The temperature of the air leaving the coil is determined by subtracting the cooling coil's effect on temperature from the dry-bulb temperature of the air entering the coil.

$$T_{\text{coil leaving}} = T_{\text{coil entering}} - \Delta T_{\text{coil}}$$

The temperature of the air entering the cooling coil is simply the mixed air dry-bulb temperature, which can be determined analytically or on the psychrometric chart using a ratio of outside air to total air quantities.

$$\begin{aligned} T_{\text{coil entering}} &= T_{\text{return}} + \left(\frac{\dot{V}_{\text{outside}}}{\dot{V}_{\text{return}} + \dot{V}_{\text{outside}}} \right) \times (T_{\text{outside}} - T_{\text{return}}) \\ &= 75°\text{F} + \left(\frac{2300 \, \frac{\text{ft}^3}{\text{min}}}{7000 \, \frac{\text{ft}^3}{\text{min}} + 2300 \, \frac{\text{ft}^3}{\text{min}}} \right) \times (90°\text{F} - 75°\text{F}) \\ &= 78.7°\text{F} \end{aligned}$$

The change in temperature across the cooling coil is determined by the sensible heat removed from the air stream. The sensible heating relationship is $\dot{q}_{\text{sensible}} = \dot{m}c_p\Delta T$.

$$\begin{aligned} \Delta T_{\text{coil}} &= \frac{\dot{q}_{\text{sensible}}}{\dot{m}c_p} = \frac{\dot{q}_{\text{sensible}}}{\dot{V}\rho c_p} \\ &= \frac{235{,}000 \, \frac{\text{Btu}}{\text{hr}}}{\left(7000 \, \frac{\text{ft}^3}{\text{min}} + 2300 \, \frac{\text{ft}^3}{\text{min}}\right)\left(0.075 \, \frac{\text{lbm}}{\text{ft}^3}\right) \times \left(0.24 \, \frac{\text{Btu}}{\text{lbm}}\right)\left(60 \, \frac{\text{min}}{\text{hr}}\right)} \\ &= 23.4°\text{F} \end{aligned}$$

The air temperature leaving the coil is

$$\begin{aligned} T_{\text{coil leaving}} &= 78.7°\text{F} - 23.4°\text{F} \\ &= 55.3°\text{F} \quad (55°\text{F}) \end{aligned}$$

The answer is (B).

67. The process from point 3 to point 4 describes the path of air leaving the cooling coil and traveling through distribution ductwork and the spaces served. The slope of the process line is called the *sensible heat ratio* and represents the relationship of the space sensible load to the space latent load. In both cases, the loads are positive, so the process involves both sensible heating and latent heating of the supply air.

The answer is (D).

68. Note the nonstandard units used by the code. The four air quantities are calculated to determine which one will govern. The negative pressure airflow requirement is

$$\begin{aligned}\dot{V}_1 &= 2610A_e\sqrt{\Delta p}\\ &= (2610)\left(1.5\ \text{ft}^2\right)\sqrt{0.05\ \text{in water}}\\ &= 875\ \text{ft}^3/\text{min}\end{aligned}$$

The gross floor area airflow requirement is

$$\begin{aligned}\dot{V}_2 &= 0.5A_{\text{gf}} = (0.5)\left(1600\ \text{ft}^2\right)\\ &= 800\ \text{ft}^3/\text{min}\end{aligned}$$

The temperature rise limitation airflow requirement is

$$\begin{aligned}\dot{V}_3 &= \frac{\sum \dot{q}}{1.08\Delta T} = \frac{15{,}000\ \frac{\text{Btu}}{\text{hr}}}{(1.08)\left(104^\circ\text{F} - 90^\circ\text{F}\right)}\\ &= 992\ \text{ft}^3/\text{min}\end{aligned}$$

The emergency purge airflow requirement is

$$\begin{aligned}\dot{V}_4 &= 100\sqrt{G} = 100\sqrt{150\ \text{lbm}}\\ &= 1225\ \text{ft}^3/\text{min}\quad \left(1200\ \text{ft}^3/\text{min}\right)\end{aligned}$$

The emergency purge requirement sets the requirement for this room.

The answer is (D).

69. From the first law of thermodynamics,

$$\begin{aligned}\dot{Q} &= \dot{m}_{\text{out}}c_p\Delta T = \dot{m}_m\Delta h\\ \dot{m}_{\text{out}} &= \frac{\dot{m}_{\text{in}}\Delta h}{c_p\Delta T}\end{aligned}$$

From saturated water tables, for a pressure of 4 psia,

$$\Delta h = h_{fg} = 1006.0\ \text{Btu/lbm}$$

For water at 65°F,

$$c_p = 0.999\ \text{Btu/lbm-}^\circ\text{F}$$

The mass flux of the exiting water is

$$\begin{aligned}\dot{m}_{\text{out}} &= \frac{\dot{m}_{\text{in}}\Delta h}{c_p\Delta T}\\ &= \frac{\left(2200\ \frac{\text{lbm}}{\text{hr}}\right)\left(1006\ \frac{\text{Btu}}{\text{lbm}}\right)}{\left(0.999\ \frac{\text{Btu}}{\text{lbm-}^\circ\text{F}}\right)\left(95^\circ\text{F} - 65^\circ\text{F}\right)}\\ &= 73{,}847\ \text{lbm/hr}\quad (74{,}000\ \text{lbm/hr})\end{aligned}$$

The answer is (C).

70. The principle is named the *Coanda effect*, after its discoverer, Henri Coanda. The lower-pressure region above an airstream traveling parallel to the ceiling causes it to adhere to the ceiling and extend the distance traveled (the throw) before reaching a defined terminal velocity.

The answer is (A).

71. The thermal energy that can be stored in the slab depends on the mass, specific heat, and the temperature difference between the slab and the room.

$$\begin{aligned}\Delta q_{\text{slab}} &= mc_p\Delta T_{\text{slab}}\\ &= V\rho c_p\Delta T_{\text{slab}}\end{aligned}$$

The temperature change of the slab is proportional to the heat lost by the slab.

$$\Delta T_{\text{slab}} = \frac{q_{\text{slab}}}{V\rho c_p}$$

Sand and gravel concrete without significant steel reinforcing that is used in construction has an approximate density of

$$\rho = 140\ \text{lbm/ft}^3$$

The concrete has an approximate specific heat of

$$c_p = 0.22\ \text{Btu/lbm-}^\circ\text{F}$$

The rate of heat loss between the slab and the room is

$$\begin{aligned}\dot{q} &= hA_{\text{slab}}\left(T_{\text{surface}} - T_{\text{air}}\right)\\ &= \left(1.4\ \frac{\text{Btu}}{\text{hr-ft}^2\text{-}^\circ\text{F}}\right)(20\ \text{ft})(30\ \text{ft})\left(82^\circ\text{F} - 60^\circ\text{F}\right)\\ &= 18{,}480\ \text{Btu/hr}\end{aligned}$$

The slab temperature change after 1 hr will be

$$\begin{aligned}\Delta T &= \frac{\Delta\dot{q}_{\text{slab}}t}{V\rho c_p}\\ &= \frac{\left(18{,}480\ \frac{\text{Btu}}{\text{hr}}\right)(1\ \text{hr})}{(20\ \text{ft})(30\ \text{ft})(0.5\ \text{ft})\left(140\ \frac{\text{lbm}}{\text{ft}^3}\right)\left(0.22\ \frac{\text{Btu}}{\text{lbm-}^\circ\text{F}}\right)}\\ &= 2.0^\circ\text{F}\quad (2^\circ\text{F})\end{aligned}$$

The answer is (B).

72. According to the ASHRAE guidelines cited, the recommended storage tank provides an hour of probable demand, appropriately factored for the facility type.

$$V_{\text{tank}} = V_{\text{probable}}F_{\text{storage factor}}$$

The probable hot-water demand is determined by adjusting the maximum demand by an industry demand factor appropriate to the facility type.

$$\dot{V}_{\text{probable}} = \dot{V}_{\text{maximum possible}} F_{\text{demand}}$$

The maximum possible demand is most easily determined in a tabular calculation format.

type of fixture	total no. of fixtures	demand per fixture (gal/hr)	total demand per fixture type (gal/hr)
lavatory	44	2	88
bathtub	22	20	440
shower	22	30	660
kitchen sink	22	10	220
dishwasher	22	15	330
maximum possible demand			1738

The maximum probable demand is

$$\begin{aligned}\dot{V}_{\text{probable}} &= \left(1738\ \frac{\text{gal}}{\text{hr}}\right)(0.30)\\ &= 521.4\ \text{gal/hr}\end{aligned}$$

The recommended storage tank is

$$\begin{aligned}V_{\text{tank}} &= V_{\text{probable}} F_{\text{storage factor}}\\ &= \dot{V}_{\text{probable}} t F_{\text{storage factor}}\\ &= \left(521.4\ \frac{\text{gal}}{\text{hr}}\right)(1\ \text{hr})(1.25)\\ &= 651.8\ \text{gal} \quad (650\ \text{gal})\end{aligned}$$

The answer is (B).

73. From the table, the solar heat gain factor for an east-facing window in May at 0800 is

$$\text{SHGF} = 218\ \text{Btu/hr-°F}$$

The shading coefficient, SC, is given as 0.87, and the overall heat transfer coefficient, U, is 1.2 Btu/hr-ft^2-°F. The instantaneous heat gain is given by the equation

$$\begin{aligned}\dot{q} &= (\text{SC})(\text{SHGF})A + UA(T_{\text{out}} - T_{\text{in}})\\ &= (0.87)\left(218\ \frac{\text{Btu}}{\text{hr-ft}^2}\right)(12\ \text{ft}^2)\\ &\quad + \left(1.2\ \frac{\text{Btu}}{\text{hr-ft}^2\text{-°F}}\right)(12\ \text{ft}^2)(42\text{°F} - 75\text{°F})\\ &= 1800.7\ \text{Btu/hr} \quad (1800\ \text{Btu/hr})\end{aligned}$$

The answer is (B).

74. The rate of energy savings will be

$$\begin{aligned}\dot{q}_{\text{savings}} &= \dot{m}_{\text{oa}} c_p \Delta T\\ &= \dot{V}_{\text{oa}} \rho_{\text{oa}} c_p (T_{\text{leaving}} - T_{\text{entering}})\end{aligned}$$

The temperature of the outside air leaving the heat exchanger will depend on both air stream quantities, entering air temperatures, and the rated heat exchanger effectiveness, ε. For sensible heat recovery, ASHRAE Standard 84 defines heat exchanger effectiveness as

$$\varepsilon = \frac{\dot{m}_{\text{supply}}(T_{\text{supply leaving}} - T_{\text{supply entering}})}{\dot{m}_{\text{min}}(T_{\text{exhaust entering}} - T_{\text{supply entering}})}$$

The supply (outside) air leaving temperature can be stated as

$$\begin{aligned}T_{\text{supply leaving}} &= T_{\text{supply entering}} + \varepsilon\left(\frac{\dot{m}_{\text{min}}}{\dot{m}_{\text{supply}}}\right)\\ &\quad \times (T_{\text{exhaust entering}} - T_{\text{supply entering}})\end{aligned}$$

$\dot{m}_{\text{min}}$ represents the mass flow rate of the lesser of the two airstreams, which is the exhaust airstream.

$$\begin{aligned}\dot{m}_{\text{min}} &= \dot{m}_{\text{exhaust}} = \dot{V}_{\text{exhaust}} \rho_{\text{exhaust}}\\ &= \frac{\dot{V}_{\text{exhaust}}}{v_{\text{exhaust}}}\end{aligned}$$

The mass flow of the supply airstream is

$$\dot{m}_{\text{supply}} = \frac{\dot{V}_{\text{supply}}}{v_{\text{supply}}}$$

From thermodynamic properties of moist air tables, at 10°F the specific volume of the supply air is

$$v_{\text{supply}} = 11.832\ \text{ft}^3/\text{lbm}$$

At 72°F the specific volume of the exhaust air is

$$v_{\text{exhaust}} = 13.400\ \text{ft}^3/\text{lbm}$$

The leaving temperature of the outside air supply is calculated as

$$\begin{aligned}&T_{\text{supply leaving}}\\ &\quad = T_{\text{supply entering}}\\ &\qquad + \varepsilon \frac{\dot{V}_{\text{exhaust}} v_{\text{supply}}}{v_{\text{exhaust}} \dot{V}_{\text{supply}}}\\ &\qquad \times (T_{\text{exhaust entering}} - T_{\text{supply entering}})\\ &\quad = 10\text{°F} + (0.64)\frac{\left(10{,}000\ \frac{\text{ft}^3}{\text{min}}\right)\left(11.832\ \frac{\text{ft}^3}{\text{lbm}}\right)}{\left(13.400\ \frac{\text{ft}^3}{\text{lbm}}\right)\left(12{,}500\ \frac{\text{ft}^3}{\text{min}}\right)}\\ &\qquad \times (72\text{°F} - 10\text{°F})\\ &\quad = 38.0\text{°F}\end{aligned}$$

The rate of energy saving is

$$\begin{aligned}\dot{q}_{\text{savings}} &= \frac{\dot{V}_{\text{supply}}}{v_{\text{supply}}} c_p \left(T_{\text{leaving}} - T_{\text{entering}}\right) \\ &= \left(\frac{12{,}500 \, \frac{\text{ft}^3}{\text{min}}}{11.832 \, \frac{\text{ft}^3}{\text{lbm}}}\right) \left(60 \, \frac{\text{min}}{\text{hr}}\right) \left(0.24 \, \frac{\text{Btu}}{\text{lbm-°F}}\right) \\ &\quad \times (38\text{°F} - 10\text{°F}) \\ &= 425{,}963 \text{ Btu/hr} \quad (430{,}000 \text{ Btu/hr})\end{aligned}$$

The answer is (A).

75. The procedure for determining the sound pressure level, L_p, at a distance d from the sound source has been standardized by ASHRAE and codified by many local governments in their noise ordinances. In the English system of units, the governing equation is

$$\begin{aligned}L_p &= L_{\text{source}} + 10 \log Q - 20 \log d - 0.5 \quad \text{[U.S.]} \\ L_p &= L_{\text{source}} + 10 \log Q - 20 \log d - 11 \quad \text{[SI]}\end{aligned}$$

Units for all terms are dB. The variable Q represents directivity and has the following values.

$Q = 1$ for uniform spherical radiation with no reflecting surfaces

$Q = 2$ for uniform hemispherical radiation with a single reflecting surface

$Q = 4$ for uniform quadrant radiation from a point at the intersection of 2 planes

With the chiller installed on the ground next to the building exterior wall, a directivity factor of $Q = 4$ is appropriate.

The sound pressure level at the property line is estimated to be

$$\begin{aligned}L_p &= 80 \text{ dB} + 10 \log 4 - 20 \log 40 \text{ ft} - 0.5 \\ &= 80 \text{ dB} + 6.02 \text{ dB} - 32.04 \text{ dB} - 0.5 \\ &= 53.47 \text{ dB} \quad (53 \text{ dB})\end{aligned}$$

The answer is (C).

76. The operative temperature, which describes human comfort in terms of temperature, radiation, and air velocity convection, is

$$T_o = AT_{\text{air}} + (1 - A)\bar{T}_r$$

For properly distributed air in a room, the velocity should be in a range of 40–60 ft/min, and

$$A = 0.5$$

The mean radiant temperature incorporates the surrounding surfaces and their respective temperatures.

$$\bar{T}_r = F_{P-1}T_1 + F_{P-2}T_2 + \ldots + F_{P-N}T_N$$

The typical mean radiant temperature is

$$\begin{aligned}\bar{T}_r &= F_{\text{glass}}T_{\text{glass}} + F_{\text{opaque}}T_{\text{opaque}} \\ &= (0.04)(35\text{°F}) + (1 - 0.04)(65\text{°F}) \\ &= 63.8\text{°F}\end{aligned}$$

The operative temperature is calculated as

$$\begin{aligned}T_o &= (0.5)(72\text{°F}) + (1 - 0.5)(63.8\text{°F}) \\ &= 67.9\text{°F} \quad (68\text{°F})\end{aligned}$$

The answer is (C).

77. The ratio of specific heats for air is

$$k = \frac{c_p}{c_v} = \frac{0.240 \, \frac{\text{Btu}}{\text{lbm-°R}}}{0.171 \, \frac{\text{Btu}}{\text{lbm-°R}}} = 1.4$$

Atmospheric pressure is 14.7 psia. The initial temperature on an absolute scale is

$$T_1 = 55\text{°F} + 460\text{°} = 515\text{°R}$$

For an isentropic process, the relationship between pressure and temperature is

$$\begin{aligned}T_2 &= T_1 \left(\frac{p_2}{p_1}\right)^{\frac{k-1}{k}} \\ &= (515\text{°R}) \left(\frac{720 \, \frac{\text{lbf}}{\text{in}^2}}{14.7 \, \frac{\text{lbf}}{\text{in}^2}}\right)^{\frac{1.4-1}{1.4}} \\ &= 1565.6\text{°R}\end{aligned}$$

For an isentropic process, work is

$$\begin{aligned}W &= -mc_v \Delta T \\ &= -(10 \text{ lbm}) \left(0.171 \, \frac{\text{Btu}}{\text{lbm-°R}}\right) (1565.6\text{°R} - 515\text{°R}) \\ &= -1796 \text{ Btu} \quad (-1800 \text{ Btu})\end{aligned}$$

The answer is (D).

78. The overall heat transfer coefficient is

$$\frac{1}{U_o} = \frac{1}{h_o} + \frac{r_o}{k}\ln\frac{r_o}{r_i} + \frac{r_o}{r_i h_i}$$

$$= \frac{1}{2.0\ \frac{\text{Btu}}{\text{hr-ft}^2\text{-°F}}} + \left(\frac{0.75\ \text{in}}{\left(12\ \frac{\text{in}}{\text{ft}}\right)\left(29\ \frac{\text{Btu}}{\text{hr-ft-°F}}\right)}\right) \times \ln\frac{0.75\ \text{in}}{0.685\ \text{in}} + \frac{0.75\ \text{in}}{(0.685\ \text{in})\left(1500\ \frac{\text{Btu}}{\text{hr-ft}^2\text{-°F}}\right)}$$

$$= 0.501\ \text{hr-ft}^2\text{-°F/Btu}$$

$$U_o = \frac{1}{0.501\ \frac{\text{hr-ft}^2\text{-°F}}{\text{Btu}}} = 2.0\ \text{Btu/hr-ft}^2\text{-°F}$$

The heat loss over the entire length of uninsulated pipe is

$$\dot{q} = U_o A_o \Delta T = U_o \pi D L \Delta T$$

$$= \left(2.0\ \frac{\text{Btu}}{\text{hr-ft}^2\text{-°F}}\right)\pi\left(\frac{1.5\ \text{in}}{12\ \frac{\text{in}}{\text{ft}}}\right)(120\ \text{ft}) \times (212\text{°F} - 60\text{°F})$$

$$= 14{,}326\ \text{Btu/hr} \quad (14{,}000\ \text{Btu/hr})$$

The answer is (A).

79. The theoretical discharge from an orifice under pressure can be calculated from Toricelli's equation using the discharge coefficient (coefficient of discharge).

$$Q = C_d A_o \sqrt{2gh}$$

However, a 1/2 in orifice with coefficient of discharge of 0.75 corresponds to the standard sprinkler, and there is no need to use theoretical methods. For a standard sprinkler, the flow for a pressure, in psig in the customary U.S. system, is given as

$$Q = K\sqrt{p} = 5.6\sqrt{p}$$

The minimum pressure of 10 psig occurs at the last sprinkler head.

$$Q = 5.6\sqrt{10\ \text{psig}} = 17.7\ \text{gal/min}$$

Disregarding the velocity head, the normal pressure at the next-to-last sprinkler is the sum of the normal pressure at the last sprinkler plus the friction loss between the two sprinklers. The friction loss can be found from the Hazen-Williams equation or (more typically) from a friction loss chart. From such a chart, the loss at 17.7 gpm for a 1 in pipe is approximately 0.1 psi/ft.

The normal pressure at the next-to-last sprinkler is

$$p = p_f + L\Delta p = 10\ \text{psig} + (10\ \text{ft})\left(0.1\ \frac{\text{psi}}{\text{ft}}\right) = 11\ \text{psig}$$

The discharge at the next-to-last sprinkler is

$$Q = 5.6\sqrt{11\ \text{psig}} = 18.6\ \text{gpm} \quad (19\ \text{gpm})$$

The answer is (C).

80. Simple payback on investment is one of the most common indexes of economic merit used when there is a need for quick decisions. The additional investment cost is divided by the annual savings attributed solely to that investment. Payback, therefore, equals incremental cost divided by annual savings.

$$t = \frac{C}{A}$$

The AFUE is a simple rating that is intended to reflect the average load-weighted efficiency of an appliance over the course of a heating season. While the prediction of energy usage can be quite complex, the estimate of savings between one scenario and another is considered to be sufficiently accurate for decision making.

The energy consumed is equal to the actual energy required (that which does not go out the furnace exhaust) divided by the average efficiency.

$$E_{\text{consumed}} = \frac{E_{\text{required}}}{\text{AFUE}}$$

The net energy required to heat the home, regardless of system losses or fuel type, is

$$E_{\text{required}} = E_{\text{consumed}}(\text{AFUE}) = \left(1500\ \frac{\text{therms}}{\text{yr}}\right)(0.78) = 1170\ \text{therms/yr}$$

The energy consumed by the more efficient furnace is

$$E_{\text{proposed}} = \frac{E_{\text{required}}}{\text{AFUE}} = \frac{1170\ \frac{\text{therms}}{\text{yr}}}{0.92} = 1272\ \text{therms/yr}$$

The annual savings is

$$\begin{aligned} E_{\text{savings}} &= E_{\text{present}} - E_{\text{proposed}} \\ &= 1500\ \frac{\text{therms}}{\text{yr}} - 1272\ \frac{\text{therms}}{\text{yr}} \\ &= 228\ \text{therms/year} \end{aligned}$$

The annual cash flow reduction is

$$\begin{aligned} A &= \left(228\ \frac{\text{therms}}{\text{yr}}\right)\left(\frac{\$0.85}{\text{therm}}\right) \\ &= \$194/\text{yr} \end{aligned}$$

The simple payback of the investment is

$$\begin{aligned} t &= \frac{\$800}{\frac{\$194}{\text{yr}}} \\ &= 4.1\ \text{yr} \quad (4\ \text{yr}) \end{aligned}$$

The answer is (B).

Solutions
Machine Design

81. Calculate the mass of the three balls.

$$\begin{aligned} m &= 3\rho V = 3\rho\tfrac{4}{3}\pi r^3 \\ &= (3)\left(0.284\ \frac{\text{lbm}}{\text{in}^3}\right)\left(\frac{4}{3}\right)\pi\left(\frac{1.00\text{ in}}{2}\right)^3 \\ &= 0.446\text{ lbm} \end{aligned}$$

Calculate the mass moment of a sphere about its centroidal axis.

$$\begin{aligned} I_c &= \tfrac{2}{5}mr^2 \\ &= \left(\frac{2}{5}\right)(0.446\text{ lbm})\left(\frac{1.00\text{ in}}{2}\right)^2 \\ &= 0.0446\text{ lbm-in}^2 \end{aligned}$$

Calculate the rotational moment of inertia using the parallel axis theorem.

$$\begin{aligned} I &= I_c + mx^2 \\ &= 0.0446\text{ lbm-in}^2 + (0.446\text{ lbm})(0.75\text{ in})^2 \\ &= 0.295\text{ lbm-in}^2 \end{aligned}$$

Calculate the natural frequency.

$$\begin{aligned} \omega &= \frac{1}{2\pi}\sqrt{\frac{kg_c}{I}} \\ &= \frac{1}{2\pi}\sqrt{\frac{\left(2.857\times 10^{-3}\ \frac{\text{in-lbf}}{\text{rad}}\right)\times\left(32.2\ \frac{\text{ft-lbm}}{\text{lbm-sec}^2}\right)\left(12\ \frac{\text{in}}{\text{ft}}\right)}{0.295\text{ lbm-in}^2}} \\ &= 0.31\text{ Hz} \end{aligned}$$

The answer is (B).

82. Resolve initial velocity into horizontal and vertical components.

$$\text{v}_{x,i} = \text{v}_i\cos\theta = \left(2000\ \frac{\text{ft}}{\text{sec}}\right)\cos 25^\circ = 1813\text{ ft/sec}$$

$$\text{v}_{y,i} = \text{v}_i\sin\theta = \left(2000\ \frac{\text{ft}}{\text{sec}}\right)\sin 25^\circ = 845\text{ ft/sec}$$

Use the acceleration of gravity and the position equation to find the elapsed time.

$$\begin{aligned} a_{y,i} = a_{y,f} &= g = -32.2\text{ ft/sec}^2 \\ s_{y,f} &= s_{y,i} + \text{v}_{y,i}t + \tfrac{1}{2}gt^2 \\ -2500\text{ ft} &= 0\text{ ft} + \left(845\ \frac{\text{ft}}{\text{sec}}\right)t \\ &\quad + \left(\frac{1}{2}\right)\left(-32.2\ \frac{\text{ft}}{\text{sec}^2}\right)t^2 \end{aligned}$$

Use the quadratic formula.

$$t = 55.3\text{ sec},\ -2.8\text{ sec}$$

Using the positive value, calculate the x displacement.

$$\begin{aligned} s_{x,f} &= s_{x,i} + \text{v}_{x,i}t \\ &= 0\text{ ft} + \left(1813\ \frac{\text{ft}}{\text{sec}}\right)(55.3\text{ sec})\left(\frac{1\text{ mi}}{5280\text{ ft}}\right) \\ &= 19\text{ mi} \end{aligned}$$

The answer is (C).

83. Calculate the reaction due to the applied moment at each bolt.

$$M = Pe = (500\text{ lbf})(8\text{ in}) = 4000\text{ in-lbf}$$

$$R_m = \frac{M}{Nr} = \frac{4000\text{ in-lbf}}{(6\text{ bolts})(1.5\text{ in})} = 444\text{ lbf/bolt}$$

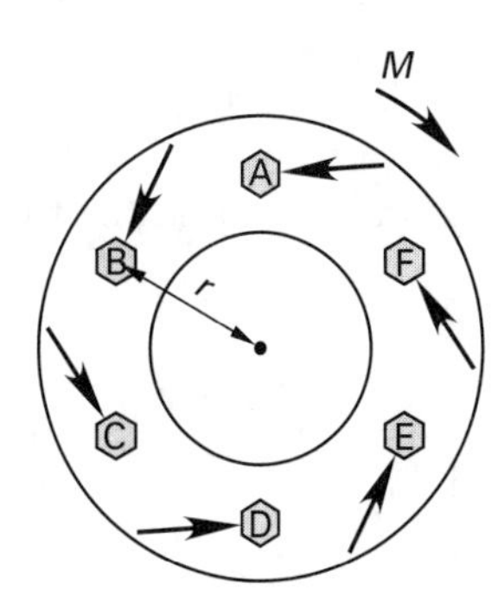

Calculate the reaction due to direct shear at each bolt.

$$R_d = \frac{P}{N} = \frac{500\text{ lbf}}{6\text{ bolts}} = 83\text{ lbf/bolt}$$

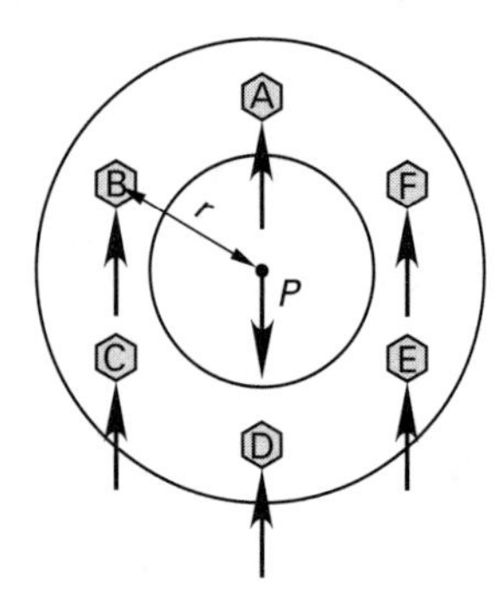

Bolts E and F have the same worst-case reaction combinations. Calculate the vector sum of reactions at bolt E.

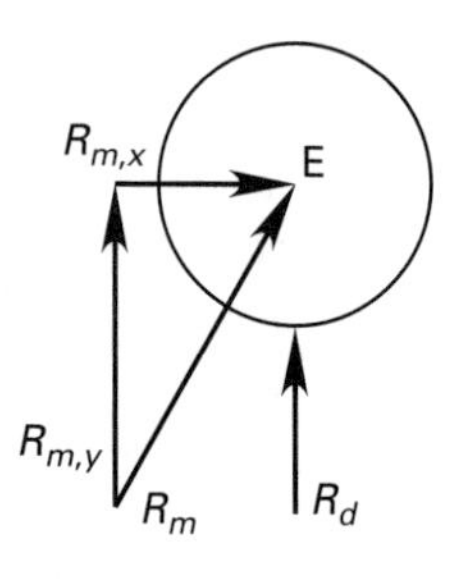

The internal angles are 30°-60°-90°. Calculate the reaction components at bolt E.

$$R_x = R_{m,x} + R_{d,x} = (444 \text{ lbf}) \sin 30° + 0 \text{ lbf} \\ = 222 \text{ lbf}$$

$$R_y = R_{m,y} + R_{d,y} = (444 \text{ lbf}) \cos 30° + 83 \text{ lbf} \\ = 468 \text{ lbf}$$

Calculate the maximum total reaction for bolt E.

$$R_{\text{E}} = \sqrt{R_x^2 + R_y^2} = \sqrt{(222 \text{ lbf})^2 + (468 \text{ lbf})^2} \\ = 518 \text{ lbf}$$

The answer is (D).

84. Find the shear modulus G, and the polar moment of inertia J. Poisson's ratio for titanium is approximately 0.34.

$$G = \frac{E}{2(1+\nu)} = \frac{15.9 \times 10^6 \ \frac{\text{lbf}}{\text{in}^2}}{(2)(1+0.34)} \\ = 5.93 \times 10^6 \text{ lbf/in}^2$$

$$J = \left(\frac{\pi}{32}\right)(D^4 - d^4) \\ = \left(\frac{\pi}{32}\right)\left((2 \text{ in})^4 - (1.75 \text{ in})^4\right) \\ = 0.650 \text{ in}^4$$

Find the maximum deflection, γ.

$$\gamma = \frac{TL}{JG} = \frac{(50{,}000 \text{ in-lbf})(2 \text{ ft})\left(12 \ \frac{\text{in}}{\text{ft}}\right)}{(0.650 \text{ in}^4)\left(5.93 \times 10^6 \ \frac{\text{lbf}}{\text{in}^2}\right)} \\ = 0.311 \text{ rad} \quad (0.31 \text{ rad})$$

The answer is (C).

85. Compute the equivalent radial load from the larger of the two values. $V_1 = 1.2$, due to outer ring rotation.

$$F_e = V_1 F_r = (1.2)(5000 \text{ lbf}) = 6000 \text{ lbf}$$

$$F_e = X V_1 F_r + Y F_t = (0.56)(1.2)(5000 \text{ lbf}) \\ + (1.31)(4000 \text{ lbf}) \\ = 8600 \text{ lbf}$$

Compute the service life.

$$\frac{L_e}{L} = \left(\frac{C}{F_e}\right)^3$$

$$L_e = L\left(\frac{C}{F_e}\right)^3 = (1{,}000{,}000 \text{ cycles})\left(\frac{5660 \text{ lbf}}{8600 \text{ lbf}}\right)^3 \\ = 285 \times 10^3 \text{ cycles} \quad (280 \times 10^3 \text{ cycles})$$

The answer is (A).

86. The velocity of point B relative to point A is the product of bar AB's length and angular velocity around point A.

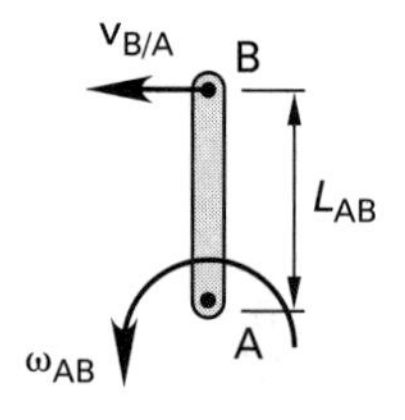

$$v_{B/A} = L_{AB}\omega_{AB}$$

Similarly, the velocity of point A relative to point O is the product of bar OA's length and angular velocity around point O.

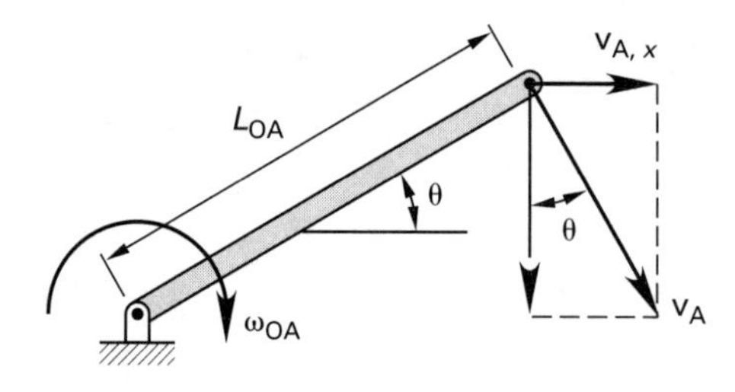

$$v_A = L_{OA}\omega_{OA}$$

The velocity at point B is equal to the vector sum of the velocity at point A and the velocity of point B relative to point A. The velocity at point B is horizontal, so the y-components of the other two velocities must sum to zero, and only the x-components need to be considered.

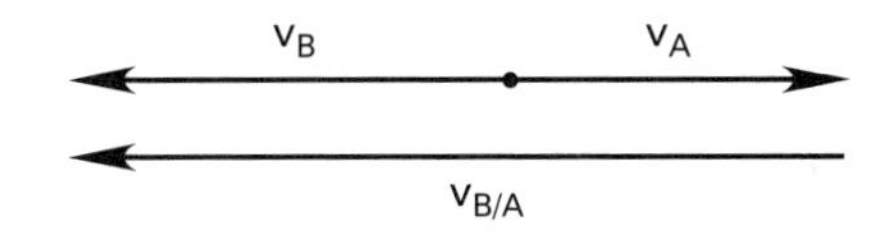

$$\begin{aligned} \mathrm{v_B} &= \mathrm{v_{A,\mathit{x}}} + \mathrm{v_{B/A,\mathit{x}}} \\ &= \mathrm{v_A} \sin\theta + \mathrm{v_{B/A,\mathit{x}}} \\ &= L_{\mathrm{OA}}\omega_{\mathrm{OA}} \sin\theta + L_{\mathrm{AB}}\omega_{\mathrm{AB}} \\ \omega_{\mathrm{AB}} &= \frac{\mathrm{v_B} - L_{\mathrm{OA}}\omega_{\mathrm{OA}}\sin\theta}{L_{\mathrm{AB}}} \\ &= \frac{5\ \frac{\text{in}}{\text{sec}} - (19\text{ in})\left(30\ \frac{\text{rad}}{\text{sec}}\right)\sin 30^\circ}{7\text{ in}} \\ &= -40\text{ rad/sec} \end{aligned}$$

The answer is (B).

87. Calculate the diametral pitch and form factor.

$$\begin{aligned} P &= \frac{n}{D} = \frac{34\text{ teeth}}{4\text{ in}} \\ &= 8.5\text{ in}^{-1} \end{aligned}$$

$$\begin{aligned} Y &= \pi y = 0.138\pi \\ &= 0.434 \end{aligned}$$

Use the Lewis equation to calculate the allowable tangential load per tooth.

$$\begin{aligned} w &= \frac{\sigma F Y}{P} = \frac{\left(40{,}000\ \frac{\text{lbf}}{\text{in}^2}\right)(1.67\text{ in})(0.434)}{8.5\text{ in}^{-1}} \\ &= 3411\text{ lbf}\quad(3400\text{ lbf}) \end{aligned}$$

The answer is (A).

88. First, find the maximum supported bending load. The distance, c, from the outermost fiber to the neutral plane is

$$\begin{aligned} c &= \frac{h}{2} = \frac{0.32\text{ in}}{2} \\ &= 0.16\text{ in} \end{aligned}$$

The maximum moment for a simply supported beam with midpoint load is

$$M_{\text{max}} = \frac{PL}{4}$$

The maximum bending load is found by rearranging the bending stress equation.

$$\begin{aligned} \sigma &= \frac{Mc}{I} = \frac{\frac{PL}{4}c}{I} \\ P &= \frac{4\sigma I}{yL} \\ &= \frac{(4)\left(29{,}000\ \frac{\text{lbf}}{\text{in}^2}\right)(8.2\times 10^{-3}\text{ in}^4)}{(0.16\text{ in})(35.0\text{ in})} \\ &= 170\text{ lbf} \end{aligned}$$

Next, find the maximum supported deflection load. The moment of inertia of the beam cross section is

$$\begin{aligned} I &= \frac{bh^3}{12} = \frac{(3.0\text{ in})(0.32\text{ in})^3}{12} \\ &= 8.19\times 10^{-3}\text{ in}^4 \end{aligned}$$

The equation for deflection of a simply supported beam with midpoint load is

$$y_{\text{max}} = \frac{PL^3}{48EI}$$

The modulus of elasticity for steel is $E = 30 \times 10^6$ lbf/in^2. The maximum supported bending load is

$$\begin{aligned} P &= \frac{48EIy_{\text{max}}}{L^3} \\ &= \frac{(48)\left(30\times 10^6\ \frac{\text{lbf}}{\text{in}^2}\right)(8.19\times 10^{-3}\text{ in}^4)(0.50\text{ in})}{(35.0\text{ in})^3} \\ &= 137.5\text{ lbf}\quad(140\text{ lbf}) \end{aligned}$$

The answer is (C).

89. Compute the mean diameter, spring index, and Wahl factor.

$$\begin{aligned} D_m &= D_o - D_w = 2.20\text{ in} - 0.225\text{ in} \\ &= 1.975\text{ in} \\ C &= \frac{D_m}{D_w} = \frac{1.975\text{ in}}{0.225\text{ in}} \\ &= 8.78 \\ K &= \frac{4C-1}{4C-4} + \frac{0.615}{C} = \frac{(4)(8.78)-1}{(4)(8.78)-4} + \frac{0.615}{8.78} \\ &= 1.17 \end{aligned}$$

Compute the maximum stress.

$$\begin{aligned} \tau_{\text{max}} &= \frac{8KFC}{\pi D_w^2} = \frac{(8)(1.17)(150\text{ lbf})(8.78)}{\pi(0.225\text{ in})^2} \\ &= 77{,}508\text{ lbf/in}^2 \end{aligned}$$

Compute the factor of safety.

$$\mathrm{FS} = \frac{\tau_{\text{allowable}}}{\tau_{\max}} = \frac{95{,}000\ \frac{\text{lbf}}{\text{in}^2}}{77{,}508\ \frac{\text{lbf}}{\text{in}^2}}$$
$$= 1.22 \quad (1.20)$$

The answer is (C).

90. Draw a free-body diagram, including reaction forces at supports A and B.

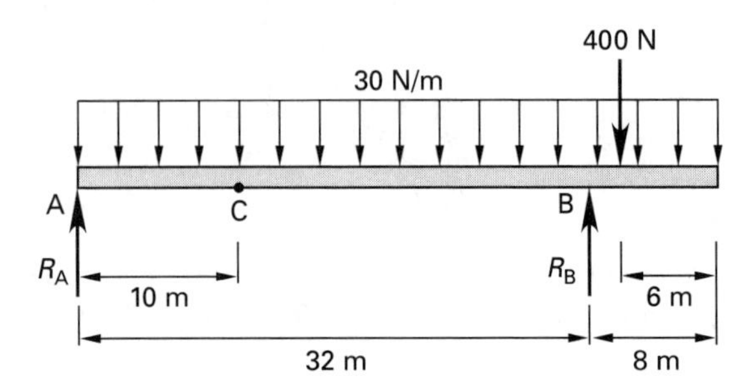

Because the beam is a rigid body at rest, both the sum of the external forces and the sum of the external moments acting on it are zero. The equilibrium equations can be used to solve for the reaction forces.

From the free-body diagram, a moment balance at support A yields

$$\sum M_{\text{A}} = 0$$
$$(32\ \text{m})R_{\text{B}} - (34\ \text{m})(400\ \text{N}) - \left(\frac{40\ \text{m}}{2}\right)(40\ \text{m})\left(30\ \frac{\text{N}}{\text{m}}\right) = 0$$
$$R_{\text{B}} = 1175\ \text{N}$$

Similarly, at support B,

$$\sum M_{\text{B}} = 0$$
$$-(32\ \text{m})(R_{\text{A}}) + (12\ \text{m})(40\ \text{m})\left(30\ \frac{\text{N}}{\text{m}}\right) - (2\ \text{m})(400\ \text{N}) = 0$$
$$R_{\text{A}} = 425\ \text{N}$$

Make a cut at point C, 10 m from the left support. Construct a free-body diagram of the section of beam to the left of the cut. (The right side could be used as an alternative.)

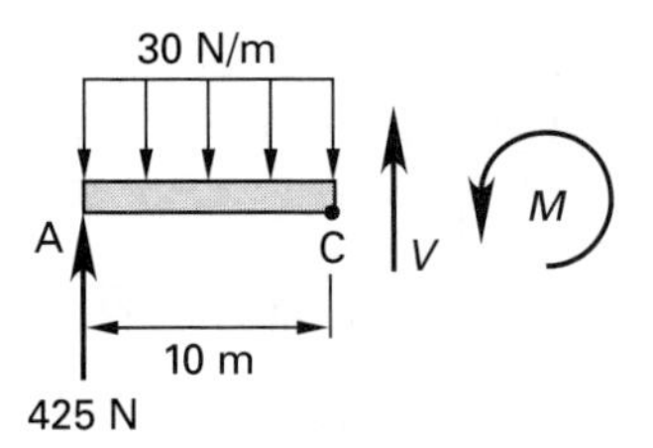

Perform a vertical force balance to determine the shear at point C.

$$\sum F_y = 0$$
$$425\ \text{N} - (10\ \text{m})\left(30\ \frac{\text{N}}{\text{m}}\right) + V = 0$$
$$V = -125\ \text{N} \quad (-130\ \text{N})$$

Whether the shear is positive or negative depends on the sign convention and free-body selected.

The answer is (A).

91. Calculate the moment of inertia.

$$I = \frac{\pi}{64}(D^4 - d^4) = \frac{\pi}{64}\left((2\ \text{in})^4 - (1.75\ \text{in})^4\right)$$
$$= 0.325\ \text{in}^4$$

Consider the shaft as a simply supported beam with two intermediate loads. Calculate deflections at A and B due to each load.

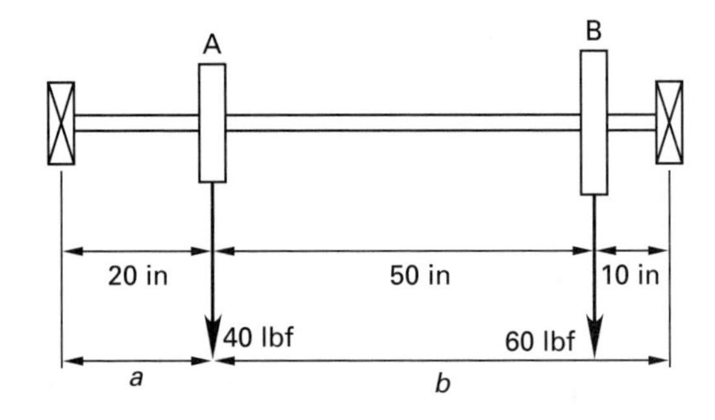

Calculate deflection at A due to the 40 lbf load.

$$y_{\text{A},40} = \frac{Pa^2b^2}{3EIL}$$
$$= \frac{(40\ \text{lbf})(20\ \text{in})^2(60\ \text{in})^2}{(3)\left(30.5 \times 10^6\ \frac{\text{lbf}}{\text{in}^2}\right)(0.325\ \text{in}^4)(80\ \text{in})}$$
$$= 0.024\ \text{in}$$

Calculate deflection at B due to the 40 lbf load.

$$y_{B,40} = \left(\frac{Pb}{6EIL}\right)\left(\frac{L}{b}(x-a)^3 + (L^2-b^2)x - x^3\right)$$

$$= \left(\frac{(40\text{ lbf})(60\text{ in})}{(6)\left(30.5\times10^6\ \frac{\text{lbf}}{\text{in}^2}\right)(0.325\text{ in}^4)(80\text{ in})}\right) \times \left(\begin{array}{l}\left(\frac{80\text{ in}}{60\text{ in}}\right)(70\text{ in} - 20\text{ in})^3 \\ \quad + \left((80\text{ in})^2 - (60\text{ in})^2\right) \\ \qquad \times (70\text{ in}) - (70\text{ in})^3\end{array}\right)$$

$$= 0.0099\text{ in} \quad (0.010\text{ in})$$

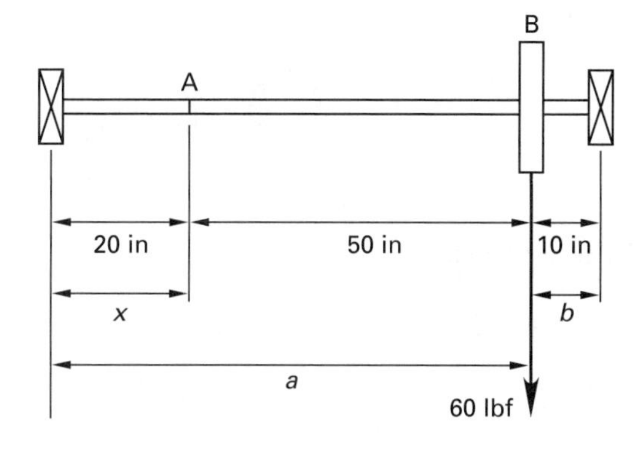

Calculate deflection at B due to the 60 lbf load.

$$y_{B,60} = \frac{Pa^2b^2}{3EIL}$$

$$= \frac{(60\text{ lbf})(70\text{ in})^2(10\text{ in})^2}{(3)\left(30.5\times10^6\ \frac{\text{lbf}}{\text{in}^2}\right)(0.325\text{ in}^4)(80\text{ in})}$$

$$= 0.012\text{ in}$$

Calculate deflection at A due to the 60 lbf load.

$$y_{A,60} = \left(\frac{Pb}{6EIL}\right)(L^2x - b^2x - x^3)$$

$$= \left(\frac{(60\text{ lbf})(10\text{ in})}{(6)\left(30.5\times10^6\ \frac{\text{lbf}}{\text{in}^2}\right)(0.325\text{ in}^4)(80\text{ in})}\right) \times \left(\begin{array}{l}(80\text{ in})^2(20\text{ in}) \\ \quad - (10\text{ in})^2(20\text{ in}) - (20\text{ in})^3\end{array}\right)$$

$$= 0.0149\text{ in} \quad (0.015\text{ in})$$

Calculate the total deflection at A and B.

$$y_{A,\text{total}} = y_{A,40} + y_{A,60} = 0.024\text{ in} + 0.015\text{ in} = 0.039\text{ in}$$

$$y_{B,\text{total}} = y_{B,40} + y_{B,60} = 0.010\text{ in} + 0.012\text{ in} = 0.022\text{ in}$$

Using the Rayleigh-Ritz equation, calculate the shaft's lowest critical speed.

$$f = \frac{1}{2\pi}\sqrt{\frac{g(W_A y_{A,\text{total}} + W_B y_{B,\text{total}})}{W_A y_{A,\text{total}}^2 + W_B y_{B,\text{total}}^2}}$$

$$= \left(\frac{1\text{ rev}}{2\pi\text{ rad}}\right) \times \sqrt{\frac{\left(384\ \frac{\text{in}}{\text{sec}^2}\right)\left((40\text{ lbf})(0.039\text{ in}) + (60\text{ lbf})(0.022\text{ in})\right)}{(40\text{ lbf})(0.039\text{ in})^2 + (60\text{ lbf})(0.022\text{ in})^2}}$$

$$= \left(17.65\ \frac{\text{rev}}{\text{sec}}\right)\left(60\ \frac{\text{sec}}{\text{min}}\right)$$

$$= 1059\text{ rpm} \quad (1100\text{ rpm})$$

The answer is (C).

92. The polar moment of inertia of the hollow shaft is

$$J = \frac{\pi D^4}{32}$$

Reformulate the expression for equivalent normal stress to solve for the inside diameter.

$$\sigma' = \frac{r}{J}\sqrt{4M^2 + 3T^2} = \frac{D}{2J}\sqrt{4M^2 + 3T^2}$$

$$= \left(\frac{16D}{\pi(D^4 - d^4)}\right)\sqrt{4M^2 + 3T^2}$$

$$d = \left(D^4 - \frac{16D}{\pi\sigma'}\sqrt{4M^2 + 3T^2}\right)^{1/4}$$

Restate the limiting stress using the maximum shear stress criterion.

$$\sigma' = S_{sy} = \frac{S_y}{2}$$

Solve for the maximum inside diameter.

$$d = \left(D^4 - \frac{16D(2)}{\pi S_y}\sqrt{4M^2 + 3T^2}\right)^{1/4}$$

$$= \left((3.0\text{ in})^4 - \frac{(16)(3.0\text{ in})(2)}{\pi\left(80{,}000\ \frac{\text{lbf}}{\text{in}^2}\right)} \times \sqrt{(4)(10{,}000\text{ in-lbf})^2 + (3)(15{,}000\text{ in-lbf})^2}\right)^{1/4}$$

$$= 2.88\text{ in} \quad (2.9\text{ in})$$

The answer is (D).

93. Calculate total shear area. The bolts are loaded in double shear.

$$A_s = (2\text{ bolts})\left(2\ \frac{\text{surfaces}}{\text{bolt}}\right)\left(\frac{\pi}{4}d^2\right) = \pi d^2$$

Using the distortion energy theory of failure, determine the ultimate shear strength.

$$\begin{aligned} S_{us} &= 0.577S_u \\ &= (0.577)\left(74{,}500\ \frac{\text{lbf}}{\text{in}^2}\right) \\ &= 42{,}990\ \text{lbf/in}^2 \end{aligned}$$

Determine the force that the pattern can support with each bolt size.

$$F = S_{us}A_s$$

size (in)	S_{us} (lbf/in²)	A_s (in²)	force supported (lbf)
1/4	42,990	0.1963	8441
3/8	42,990	0.4418	18,992
1/2	42,990	0.7854	33,764
5/8	42,990	1.2272	52,757

The answer is (B).

94. Calculate the bolt force at the proof stress.

$$\begin{aligned} F = \sigma_{\text{proof}}A_t &= \left(105{,}000\ \frac{\text{lbf}}{\text{in}^2}\right)(0.606\ \text{in}^2) \\ &= 63{,}630\ \text{lbf} \end{aligned}$$

Determine the torque required to tighten the nut.

$$\begin{aligned} \alpha &= 30^\circ \\ \theta &= 2.48^\circ \\ r_t &= \frac{d_t}{2} = \frac{0.9188\ \text{in}}{2} = 0.4594\ \text{in} \\ T &= \left(\frac{f_c r_c}{d_{\text{bolt}}} + \left(\frac{r_t}{d_{\text{bolt}}}\right)\left(\frac{\tan\theta + f_t\sec\alpha}{1 - f_t\tan\theta\sec\alpha}\right)\right)d_{\text{bolt}}F \\ &= \left(\frac{(0.15)(0.625\ \text{in})}{1.00\ \text{in}} + \left(\frac{0.4594\ \text{in}}{1.00\ \text{in}}\right)\left(\frac{\tan 2.48^\circ + 0.15\sec 30^\circ}{1 - 0.15\tan 2.48^\circ\sec 30^\circ}\right) \times (1.00\ \text{in})(63{,}630\ \text{lbf})\right)\left(\frac{1\ \text{ft}}{12\ \text{in}}\right) \\ &= 1029\ \text{ft-lbf}\quad (1000\ \text{ft-lbf}) \end{aligned}$$

The answer is (A).

95. The problem can be solved analytically using the Freudenstein equation, a fundamental kinematic identity.

$$R_1\cos\phi - R_2\cos\theta + R_3 = \cos(\phi - \theta)$$

Simplify using a trigonometric identity, and solve for the coefficients.

$$\begin{aligned} \cos(\phi-\theta) &= \cos\phi\cos\theta + \sin\phi\sin\theta \\ R_1 &= \frac{a}{d} = \frac{5\ \text{in}}{4\ \text{in}} = 5/4 \\ R_2 &= \frac{a}{b} = \frac{5\ \text{in}}{2\ \text{in}} = 5/2 \\ R_3 &= \frac{b^2 - c^2 + d^2 + a^2}{2bd} \\ &= \frac{(2\ \text{in})^2 - (3\ \text{in})^2 + (4\ \text{in})^2 + (5\ \text{in})^2}{(2)(2\ \text{in})(4\ \text{in})} \\ &= 36/16 \end{aligned}$$

Substituting and simplifying,

$$\begin{aligned} \frac{5}{4}\cos 120^\circ - \frac{5}{2}\cos\theta + \frac{36}{16} &= \cos 120^\circ\cos\theta + \sin 120^\circ\sin\theta \\ 1 &= \frac{16}{13}\cos\theta + \frac{4\sqrt{3}}{13}\sin\theta \end{aligned}$$

Solve for the output angle by selecting trial θ values and tabulating results.

θ (deg)	$1 - \frac{4\sqrt{3}}{13}\sin\theta$	$\frac{16}{13}\cos\theta$
55	0.563	0.706
65	0.517	0.520
65.2	**0.516**	**0.516**
75	0.485	0.318
85	0.469	0.107

The answer is (B).

96. Calculate the required thermal expansion coefficient.

$$\begin{aligned} \frac{x_2 - x_1}{(T_2 - T_1)x_1} &= \frac{10.045\ \text{in} - 10.000\ \text{in}}{(80^\circ\text{C} - 25^\circ\text{C})(10.000\ \text{in})} \\ &= 81.8 \times 10^{-6}\ \text{in/in-}^\circ\text{C} \end{aligned}$$

Material I has a higher coefficient of thermal expansion and is disqualified. Calculate the required specific gravity.

$$\begin{aligned} \text{SG}_{\text{req}} &\le 0.5(\text{SG}_{\text{al}}) \\ &\le (0.5)(2.71) \\ &= 1.355 \end{aligned}$$

Material II has a higher specific gravity than required and is disqualified. Calculate the operating temperature in degrees Farenheit.

$$T_{^\circ F} = \tfrac{9}{5}T_{^\circ C} + 32^\circ = \left(\frac{9}{5}\right)(80^\circ C) + 32^\circ$$
$$= 176^\circ F$$

Material III has a lower maximum temperature and is disqualified. Material IV is the only structural polymer of this group that meets the design engineers' requirements.

The answer is (D).

97. Find the maximum principal stress.

$$\sigma_1 = \frac{\sigma_x + \sigma_y}{2} + \sqrt{\left(\frac{\sigma_x - \sigma_y}{2}\right)^2 + \tau_{xy}^2}$$
$$= \frac{10{,}000\ \frac{\text{lbf}}{\text{in}^2} + 15{,}000\ \frac{\text{lbf}}{\text{in}^2}}{2} + \sqrt{\left(\frac{10{,}000\ \frac{\text{lbf}}{\text{in}^2} - 15{,}000\ \frac{\text{lbf}}{\text{in}^2}}{2}\right)^2 + \left(5000\ \frac{\text{lbf}}{\text{in}^2}\right)^2}$$
$$= 18{,}090\ \text{lbf/in}^2$$

Find the maximum shear stress.

$$\tau_{\max} = \sqrt{\left(\frac{\sigma_x - \sigma_y}{2}\right)^2 + \tau_{xy}^2}$$
$$= \sqrt{\left(\frac{10{,}000\ \frac{\text{lbf}}{\text{in}^2} - 15{,}000\ \frac{\text{lbf}}{\text{in}^2}}{2}\right)^2 + \left(5000\ \frac{\text{lbf}}{\text{in}^2}\right)^2}$$
$$= 5590\ \text{lbf/in}^2$$

Find the factor of safety for normal stress.

$$\text{FS} = \frac{\sigma_{\text{yield}}}{\sigma_1} = \frac{45{,}000\ \frac{\text{lbf}}{\text{in}^2}}{18{,}090\ \frac{\text{lbf}}{\text{in}^2}}$$
$$= 2.49 \quad (2.5)$$

Find the factor of safety for shear stress.

$$\text{FS} = \frac{\tau_{\text{yield}}}{\tau_{\max}} = \frac{10{,}000\ \frac{\text{lbf}}{\text{in}^2}}{5590\ \frac{\text{lbf}}{\text{in}^2}}$$
$$= 1.79 \quad (1.8)$$

The answer is (A).

98. Solve using the equivalent uniform annual cost method.

Calculate the annual cost of the upgrade alternative.

$$\begin{aligned}\text{EUAC}_u &= \text{annual maintenance} \\ &\quad + (\text{initial cost} + \text{present salvage value}) \\ &\quad \times (\text{capital recovery factor}) \\ &\quad - (\text{future salvage value}) \\ &\quad \times (\text{uniform series sinking fund factor}) \\ &= \$500 + (\$9000 + \$13{,}000)(A/P, 8\%, 20) \\ &\quad - (\$10{,}000)(A/F, 8\%, 20) \\ &= \$500 + (\$9000 + \$13{,}000)(0.1019) \\ &\quad - (\$10{,}000)(0.0219) \\ &= \$2523\end{aligned}$$

Calculate the annual cost of the replacement alternative.

$$\begin{aligned}\text{EUAC}_r &= \text{annual maintenance} \\ &\quad + (\text{initial cost})(\text{capital recovery factor}) \\ &\quad - (\text{future salvage value}) \\ &\quad \times (\text{uniform series sinking fund factor}) \\ &= \$100 + (\$40{,}000)(A/P, 8\%, 25) \\ &\quad - \$15{,}000(A/F, 8\%, 25) \\ &= \$100 + (\$40{,}000)(0.0937) \\ &\quad - (\$15{,}000)(0.0137) \\ &= \$3642\end{aligned}$$

The difference in annual costs is

$$\begin{aligned}\text{EUAC}_r - \text{EUAC}_u &= \$3642 - \$2523 \\ &= \$1119 \quad (\$1100)\end{aligned}$$

The upgrade alternative is more economical.

The answer is (C).

99. Calculate the average radius.

$$r_c = \frac{r_o + r_i}{2} = \frac{18\ \text{in} + 16\ \text{in}}{2}$$
$$= 17\ \text{in}$$

Calculate the design speed.

$$\sigma_t = \frac{S_u}{\text{FS}} = \frac{\rho v_c^2}{g_c}$$

$$v_c = \sqrt{\frac{g_c S_u}{\text{FS}\rho}}$$

$$= \sqrt{\frac{\left(384 \ \frac{\text{in-lbm}}{\text{lbf-sec}^2}\right)\left(100{,}000 \ \frac{\text{lbf}}{\text{in}^2}\right)}{(10)\left(0.26 \ \frac{\text{lbm}}{\text{in}^3}\right)}}$$

$$= 3843 \text{ in/sec}$$

Convert to rpm.

$$n = v_c\left(\frac{1 \text{ rev}}{2\pi r_c}\right)$$

$$= \left(3843 \ \frac{\text{in}}{\text{sec}}\right)\left(\frac{1 \text{ rev}}{2\pi(17 \text{ in})}\right)\left(60 \ \frac{\text{sec}}{\text{min}}\right)$$

$$= 2159 \text{ rev/min} \quad (2200 \text{ rpm})$$

The answer is (B).

100. Tabulate fatigue life versus completely reversed stress from the given table.

completely reversed stress (lbf/in^2)	fatigue life (cycles)
±80,000	5×10^4
±50,000	3×10^5
±30,000	1×10^6

Apply the linear cumulative damage (Palmgren-Miner) rule.

$$\sum_{j=1}^{k} \frac{n_j}{N_j} = 1$$

Calculate the lifetime fraction consumed by one 9 min cycle.

$$L_{\text{fraction,1 cycle}} = \frac{1}{5 \times 10^4} + \frac{1}{3 \times 10^5} + \frac{1}{1 \times 10^6}$$

$$= 2.43 \times 10^{-5}$$

Calculate the total number of cycles per lifetime.

$$N_{\text{total}} = \frac{1}{L_{\text{fraction}}} = \frac{1 \text{ cycle}}{2.43 \times 10^{-5}}$$

$$= 41{,}152 \text{ cycles}$$

Calculate the lifetime in hours.

$$t_{\text{total}} = (41{,}152 \text{ cycles})\left(9 \ \frac{\text{min}}{\text{cycle}}\right)\left(\frac{1 \text{ hr}}{60 \text{ min}}\right)$$

$$= 6173 \text{ hr} \quad (6200 \text{ hr})$$

The answer is (B).

101. Calculate the effective length. For fixed-free end conditions,

$$C = 2.1$$

$$L_e = CL = (2.1)(2 \text{ in})$$

$$= 4.2 \text{ in}$$

Calculate the section area, minimum moment of inertia, minimum radius of gyration, and slenderness ratio.

$$A = bh = (1.0 \text{ in})(0.7 \text{ in})$$

$$= 0.7 \text{ in}^2$$

$$I_{\text{min}} = \tfrac{1}{12}bh^3 = \left(\frac{1}{12}\right)(1.0 \text{ in})(0.7 \text{ in})^3$$

$$= 0.0286 \text{ in}^4$$

$$r_{\text{min}} = \sqrt{\frac{I_{\text{min}}}{A}} = \sqrt{\frac{0.0286 \text{ in}^4}{0.7 \text{ in}^2}}$$

$$= 0.2 \text{ in}$$

$$\text{SR} = \frac{L_e}{r_{\text{min}}} = \frac{4.2 \text{ in}}{0.2 \text{ in}}$$

$$= 21$$

Calculate the critical slenderness ratio. If the slenderness ratio is less than the critical slenderness ratio, the column is not long, and the Johnson formula is used to calculate the critical load.

$$\text{SR}_{\text{critical}} = \sqrt{\frac{2\pi^2 E}{S_y}} = \sqrt{\frac{2\pi^2\left(420{,}000 \ \frac{\text{lbf}}{\text{in}^2}\right)}{10{,}000 \ \frac{\text{lbf}}{\text{in}^2}}}$$

$$= 28.8$$

Since 21 < 28.8, use the Johnson formula to calculate the critical load.

$$P_{\text{critical}} = AS_y\left(1 - \frac{S_y(\text{SR})^2}{4\pi^2 E}\right)$$

$$= (0.7 \text{ in}^2)\left(10{,}000 \ \frac{\text{lbf}}{\text{in}^2}\right) \times \left(1 - \frac{\left(10{,}000 \ \frac{\text{lbf}}{\text{in}^2}\right)(21)^2}{4\pi^2\left(420{,}000 \ \frac{\text{lbf}}{\text{in}^2}\right)}\right)$$

$$= 5138 \text{ lbf} \quad (5100 \text{ lbf})$$

The answer is (A).

102. For processes conforming to the standard normal distribution, probability of a given outcome is calculated using the integral of the standard normal distribution equation. The standard normal distribution equation is

$$f(x) = \frac{1}{2\pi} e^{-\frac{1}{2}x^2}$$

$$\int_{-\infty}^{+\infty} f(x) = \pi$$

Integration across the entire domain equals π as shown, but solution for intermediate domains is difficult. Statistics texts provide tabulated solutions. The z-transform simplifies probability calculation by normalizing standard deviations and setting the mean to zero. The z-transform curve is shown.

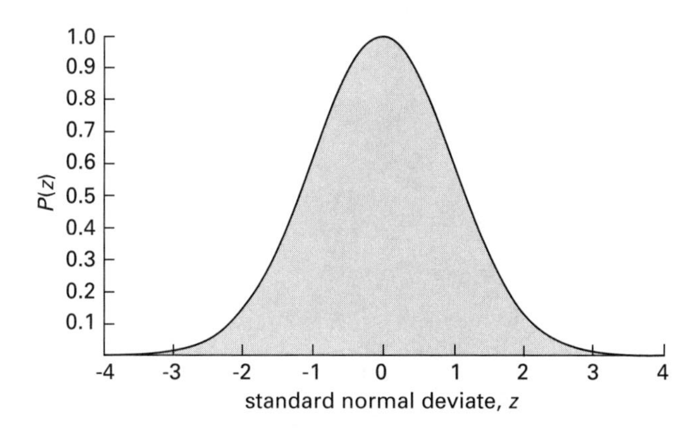

The following table provides z-transform solution for integer values of z.

	A	B	C	D	E
	area	area	A + B	1 − C	
z	$-\infty$ to z	z to $+\infty$	reject	accept	percent
0.00	0.5000	0.5000	1.0000	0.0000	0.0%
1.00	0.1587	0.1587	0.3174	0.6826	68.3%
2.00	0.0228	0.0228	0.0456	0.9544	95.4%
3.00	0.0014	0.0014	0.0028	0.9972	99.7%
4.00	3.2×10^{-5}	3.2×10^{-5}	6.4×10^{-5}	0.9999	99.9%

The probability of an accepted process outcome is represented by column D. For example, for $-1 < z < 1$ (corresponding to $\pm 1\sigma$) the probability of success is 0.6826. For $-2 < z < 2$ ($\pm 2\sigma$) the probability of success is 0.9544. For $-3 < z < 3$ ($\pm 3\sigma$) the probability of success is 0.9972.

The answer is (C).

103. Evaluate each joint, calculating the effective allowable stress, geometry after corrosion, and allowable pressure. Verify that the results meet code limitations.

Evaluate the type 1 spot radiographed longitudinal joint. Calculate the effective allowable stress.

$$\begin{aligned} \sigma_{850^\circ\text{F}} &= 8.7 \times 10^3 \text{ lbf/in}^2 \\ SE_\text{long} &= \sigma_{850^\circ\text{F}} E \\ &= \left(8.7 \times 10^3 \ \frac{\text{lbf}}{\text{in}^2}\right)(0.85) \\ &= 7395 \text{ lbf/in}^2 \end{aligned}$$

Calculate the relevant geometry (shell thickness and inside radius) after corrosion.

$$\begin{aligned} t_c &= t - \text{CA} \\ &= 0.625 \text{ in} - \frac{1}{16} \text{ in} \\ &= 0.5625 \text{ in} \\ R_c &= \frac{\text{OD}}{2} - t + \text{CA} \\ &= \frac{24 \text{ in}}{2} - 0.625 \text{ in} + \frac{1}{16} \text{ in} \\ &= 11.44 \text{ in} \end{aligned}$$

Calculate the maximum allowed pressure.

$$\begin{aligned} p_\text{max} &= \frac{SE_\text{long} t_c}{R_c + 0.6 t_c} \\ &= \frac{\left(7395 \ \frac{\text{lbf}}{\text{in}^2}\right)(0.5625 \text{ in})}{11.44 \text{ in} + (0.6)(0.5625 \text{ in})} \\ &= 353 \text{ lbf/in}^2 \end{aligned}$$

Verify that results meet code limitations.

$$\begin{aligned} p_\text{max} &= 353 \ \frac{\text{lbf}}{\text{in}^2} \le 0.385 SE \\ &= 353 \ \frac{\text{lbf}}{\text{in}^2} \le (0.385)\left(7395 \ \frac{\text{lbf}}{\text{in}^2}\right) \\ &= 353 \ \frac{\text{lbf}}{\text{in}^2} \le 2847 \ \frac{\text{lbf}}{\text{in}^2} \quad [\text{ok}] \\ t_c &= 0.562 \text{ in} \le 0.5 R_c \\ &= 0.562 \text{ in} \le (0.5)(11.44 \text{ in}) \\ &= 0.562 \text{ in} \le 5.72 \text{ in} \quad [\text{ok}] \end{aligned}$$

Evaluate the type 2 spot radiographed circumferential joint. Calculate the effective allowable stress.

$$\begin{aligned} SE_\text{circ} &= \sigma_{850^\circ\text{F}} E \\ &= \left(8.7 \times 10^3 \ \frac{\text{lbf}}{\text{in}^2}\right)(0.80) \\ &= 6960 \text{ lbf/in}^2 \end{aligned}$$

Calculate the relevant geometry after corrosion. (Use the shell thickness and inside radius calculated for the longitudinal joint.)

$$t_c = 0.562 \text{ in}$$
$$R_c = 11.44 \text{ in}$$

Calculate the maximum allowable pressure.

$$\begin{aligned} p_{\max} &= \frac{2SE_{\text{circ}}t_c}{R_c - 0.4t_c} \\ &= \frac{(2)\left(6960 \ \frac{\text{lbf}}{\text{in}^2}\right)(0.5625 \text{ in})}{11.44 \text{ in} - (0.4)(0.5625 \text{ in})} \\ &= 699 \text{ lbf/in}^2 \end{aligned}$$

Verify that results meet code limitations.

$$\begin{aligned} p_{\max} &= 699 \ \frac{\text{lbf}}{\text{in}^2} \le 1.25SE \\ &= 699 \ \frac{\text{lbf}}{\text{in}^2} \le (1.25)\left(6960 \ \frac{\text{lbf}}{\text{in}^2}\right) \\ &= 699 \ \frac{\text{lbf}}{\text{in}^2} \le 8700 \ \frac{\text{lbf}}{\text{in}^2} \quad \text{[ok]} \\ t_c &= 0.562 \text{ in} \le 0.5R_c \\ &= 0.562 \text{ in} \le (0.5)(11.44 \text{ in}) \\ &= 0.562 \text{ in} \le 5.72 \text{ in} \quad \text{[ok]} \end{aligned}$$

Evaluate the elliptical head. Calculate the effective allowable stress.

$$\begin{aligned} SE_{\text{ehead}} &= \sigma_{850^\circ\text{F}}E \\ &= \left(8.7 \times 10^3 \ \frac{\text{lbf}}{\text{in}^2}\right)(1.0) \\ &= 8700 \text{ lbf/in}^2 \end{aligned}$$

Calculate the relevant geometry (shell thickness, shell diameter, and K-factor) after corrosion. (Use the shell thickness calculated for the longitudinal joint).

$$\begin{aligned} t_c &= 0.562 \text{ in} \\ D_c &= 2R_c = (2)(11.44 \text{ in}) = 22.88 \text{ in} \\ h &= 6 \text{ in} \\ K &= \left(\frac{1}{6}\right)\left(2 + \left(\frac{D_c}{2h}\right)^2\right) \\ &= \left(\frac{1}{6}\right)\left(2 + \left(\frac{22.88 \text{ in}}{(2)(6 \text{ in})}\right)^2\right) \\ &= 0.939 \end{aligned}$$

Calculate maximum allowable pressure.

$$\begin{aligned} p_{\max} &= \frac{2SE_{\text{ehead}}t_c}{KD_c + 0.2t_c} \\ &= \frac{(2)\left(8700 \ \frac{\text{lbf}}{\text{in}^2}\right)(0.5625 \text{ in})}{(0.939)(22.88 \text{ in}) + (0.2)(0.5625 \text{ in})} \\ &= 453 \text{ lbf/in}^2 \end{aligned}$$

Code limitations were satisfied by the calculation of K.

Evaluate the hemispherical head. Calculate the effective allowable stress.

$$\begin{aligned} SE_{\text{hhead}} &= \sigma_{850^\circ\text{F}}E \\ &= \left(8.7 \times 10^3 \ \frac{\text{lbf}}{\text{in}^2}\right)(0.80) \\ &= 6960 \text{ lbf/in}^2 \end{aligned}$$

Calculate the relevant geometry (head thickness and inside radius) after corrosion.

$$\begin{aligned} t_c &= t - \text{CA} \\ &= 0.5 \text{ in} - \frac{1}{16} \text{ in} \\ &= 0.4375 \text{ in} \\ R_c &= R - t - \text{CA} \\ &= \frac{24 \text{ in}}{2} - 0.5 \text{ in} - \frac{1}{16} \text{ in} \\ &= 11.438 \text{ in} \end{aligned}$$

Calculate the maximum allowable pressure.

$$\begin{aligned} p_{\max} &= \frac{2SE_{\text{hhead}}t_c}{R_c + 0.2t_c} \\ &= \frac{(2)\left(6960 \ \frac{\text{lbf}}{\text{in}^2}\right)(0.4375 \text{ in})}{11.438 \text{ in} + (0.2)(0.4375 \text{ in})} \\ &= 528 \text{ lbf/in}^2 \end{aligned}$$

Verify that results meet code limitations.

$$\begin{aligned} p_{\max} &= 528 \ \frac{\text{lbf}}{\text{in}^2} \le 0.665SE \\ &= 528 \ \frac{\text{lbf}}{\text{in}^2} \le (0.665)\left(6960 \ \frac{\text{lbf}}{\text{in}^2}\right) \\ &= 528 \ \frac{\text{lbf}}{\text{in}^2} \le 4628 \ \frac{\text{lbf}}{\text{in}^2} \quad \text{[ok]} \\ t_c &= 0.4375 \text{ in} \le 0.356R_c \\ &= 0.4375 \text{ in} \le (0.356)(11.125 \text{ in}) \\ &= 0.4375 \text{ in} \le 3.96 \text{ in} \quad \text{[ok]} \end{aligned}$$

The lowest allowable stress, 353 lbf/in^2 (350 lbf/in^2), occurs on the longitudinal weld.

The answer is (A).

104. Calculate the impact factor.

$$\begin{aligned} f_i &= 1+\sqrt{1+\frac{2h}{\delta_{st}}} \\ &= 1+\sqrt{1+\frac{(2)(6\text{ ft})\left(12\ \frac{\text{in}}{\text{ft}}\right)}{5\times 10^{-3}\text{ in}}} \\ &= 170.7 \end{aligned}$$

Use the impact factor, effective force, failure deflection, and factor of safety to find the allowable weight.

$$\begin{aligned} \frac{F_e}{N} &= W f_i \\ W &= \frac{F_e}{N f_i} \\ &= \frac{1250\text{ lbf}}{(6)(170.7)} \\ &= 1.22\text{ lbf} \quad (1.2\text{ lbf}) \end{aligned}$$

The answer is (D).

105. The cycloidal profile provides zero acceleration at start of rise and end of return. The harmonic and parabolic profiles impose instantaneous accelerations upon start of rise and end of return. There is no profile named velocity derivative.

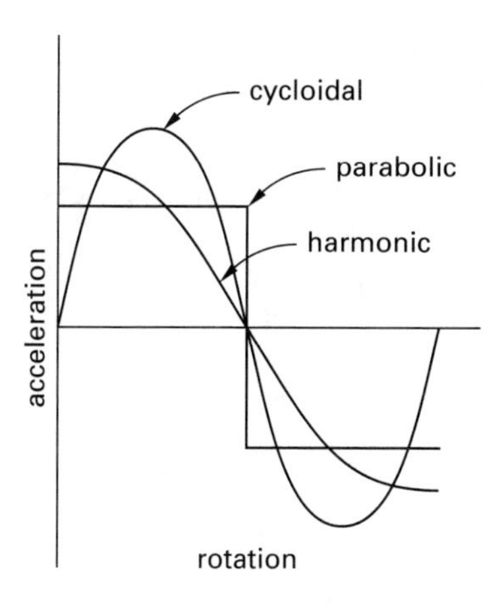

The answer is (B).

106. The thickness of the beam is doubled. The thickness ratio, t_2/t_1, is 2. The cross section is rectangular, so the mass ratio, m_2/m_1, is also 2. The moment of inertia ratio is

$$\begin{aligned} \frac{I_2}{I_1} &= \frac{\frac{1}{12}bt_2^3}{\frac{1}{12}bt_1^3} = \left(\frac{t_2}{t_1}\right)^3 \\ &= (2)^3 \\ &= 8 \end{aligned}$$

The stiffness constant for a cantilever beam is

$$k = \frac{P}{\delta} = \frac{3EI}{L^3}$$

The stiffness ratio is

$$\begin{aligned} \frac{k_2}{k_1} &= \frac{\frac{3EI_2}{L^3}}{\frac{3EI_1}{L^3}} = \frac{I_2}{I_1} \\ &= 8 \end{aligned}$$

The fundamental natural frequency ratio is

$$\begin{aligned} \frac{\omega_{n,2}}{\omega_{n,1}} &= \frac{\sqrt{\frac{k_2}{m_2}}}{\sqrt{\frac{k_1}{m_1}}} = \sqrt{\frac{\frac{k_2}{k_1}}{\frac{m_2}{m_1}}} = \sqrt{\frac{8}{2}} \\ &= 2 \end{aligned}$$

The answer is (C).

107. Although a third party often assists in design, evaluation, testing, and regulatory interpretation, the CE marking is not a third-party approval mark. It is a self-declaration under the supplier's own responsibility that the product conforms to European Union requirements. By affixing the CE marking to a product, the responsible company official declares successful completion of appropriate conformity assessment procedures.

The answer is (B).

108. The moment about a fulcrum, such as the end of a beam, is the cross product of the force applied and its associated position vector. When multiple forces exist, the net moment is the sum of all cross products.

The moment created by the distributed load is

$$\begin{aligned} M_{\text{distributed}} &= Fd = wL\left(d+\frac{L}{2}\right) \\ &= \left(20\ \frac{\text{lbf}}{\text{ft}}\right)(8\text{ ft})\left(7\text{ ft}+\frac{8\text{ ft}}{2}\right) \\ &= 1760\text{ ft-lbf} \end{aligned}$$

The moment created by the concentrated load is

$$\begin{aligned} M_{\text{concentrated}} &= Fd \\ &= (100\text{ lbf})(15\text{ ft}) \\ &= 1500\text{ ft-lbf} \end{aligned}$$

The net moment is

$$\begin{aligned} M_{\text{net}} &= M_{\text{distributed}} + M_{\text{concentrated}} \\ &= 1760 \text{ ft-lbf} + 1500 \text{ ft-lbf} \\ &= 3260 \text{ ft-lbf} \quad (3300 \text{ ft-lbf}) \end{aligned}$$

The answer is (C).

109. The moment of inertia for the rod is

$$\begin{aligned} I &= \frac{\pi}{64} d^4 = \frac{\pi}{64} (2 \text{ in})^4 \\ &= 0.7854 \text{ in}^4 \end{aligned}$$

The modulus of elasticity for steel is 30×10^6 psi. The stiffness constant for a cantilever beam is

$$\begin{aligned} k &= \frac{P}{\delta} = \frac{3EI}{L^3} \\ &= \frac{(3)\left(30 \times 10^6 \ \frac{\text{lbf}}{\text{in}^2}\right)(0.7854 \text{ in}^4)}{(36 \text{ in})^3} \\ &= 1515 \text{ lbf/in} \end{aligned}$$

The natural frequency is

$$\begin{aligned} \omega_n &= \sqrt{\frac{kg_c}{m}} \\ &= \sqrt{\frac{\left(1515 \ \frac{\text{lbf}}{\text{in}}\right)\left(32.2 \ \frac{\text{lbm-ft}}{\text{lbf-sec}^2}\right)\left(12 \ \frac{\text{in}}{\text{ft}}\right)}{50 \text{ lbm}}} \\ &= 108.2 \text{ rad/sec} \end{aligned}$$

The damping frequency is

$$\begin{aligned} \zeta &= \frac{Cg_c}{2m\omega} \\ &= \frac{\left(1.0 \ \frac{\text{lbf-sec}}{\text{in}}\right)\left(32.2 \ \frac{\text{lbm-ft}}{\text{lbf-sec}^2}\right)\left(12 \ \frac{\text{in}}{\text{ft}}\right)}{(2)(50 \text{ lbm})\left(108.2 \ \frac{\text{rad}}{\text{sec}}\right)} \\ &= 0.0357 \quad (0.036) \end{aligned}$$

The answer is (B).

110. Draw a free-body diagram for the given dynamic conditions. Note that the pulley system doubles the pull on the crate toward the top of the ramp.

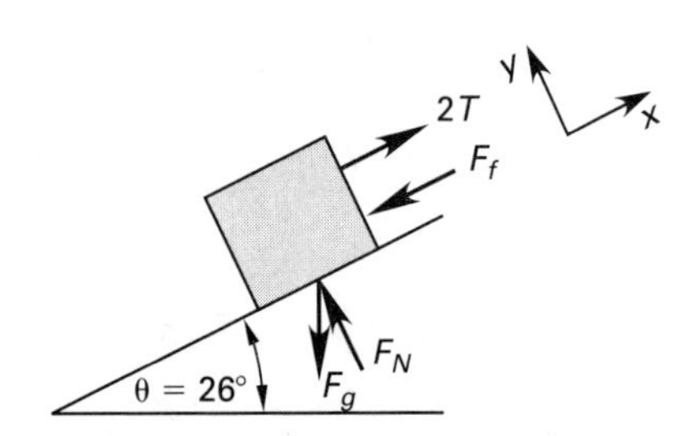

Determine the normal force acting on the crate by using a force balance in the direction perpendicular to the plane of motion (y-direction).

$$\sum F_y = 0$$

$$F_N - F_g \cos\theta = 0$$

$$\begin{aligned} F_N &= F_g \cos\theta = mg\cos\theta \\ &= (500 \text{ kg})\left(9.81 \ \frac{\text{m}}{\text{s}^2}\right)\cos 26^\circ \\ &= 4409 \text{ N} \end{aligned}$$

Because the velocity of the tractor is constant, the crate's acceleration up the ramp is zero. A new force balance in the direction of motion (x-direction) yields

$$\sum F_x = ma_x = 0$$

$$2T - F_f - F_g \sin\theta = 0$$

Solve for the tension in the cable, T.

$$\begin{aligned} T &= \tfrac{1}{2}(F_f + F_g \sin\theta) \\ &= \tfrac{1}{2}(\mu_k F_N + mg\sin\theta) \\ &= \left(\frac{1}{2}\right)\left(\begin{array}{l} (0.24)(4409 \text{ N}) \\ \quad + (500 \text{ kg})\left(9.81 \ \frac{\text{m}}{\text{s}^2}\right)\sin 26^\circ \end{array}\right) \\ &= 1604 \text{ N} \quad (1600 \text{ N}) \end{aligned}$$

The answer is (C).

111. Cathodic protection is a technique to prevent metal surface corrosion by making that surface the cathode of an electrochemical cell. For example, surfaces can be cathodically protected through the application of either galvanic anodes or impressed current.

Anodic protection is the application of direct current to shift corrosion potential to the passive zone in active-passive metals exposed to strong alkaline or acidic environments.

Passivation is the process of cleaning metal surfaces and forming a corrosion-resistant film, particularly on stainless steel, but also on aluminum, copper, and other metals. For stainless steel, the surface removal of iron or iron compounds is accomplished by means of a chemical dissolution, typically an acid treatment. Treatment with a mild nitric acid solution or other oxidant forms a protective passive film.

In the process of galvanization, the metal surface is coated with protective zinc by dipping or electrodeposition.

The answer is (A).

112. The radius of the cylinder in feet is

$$\begin{aligned} r &= (15 \text{ in})\left(\frac{1 \text{ ft}}{12 \text{ in}}\right) \\ &= 1.25 \text{ in} \end{aligned}$$

The mass moment of inertia of the cylinder is

$$\begin{aligned} I_{\text{cylinder}} &= \tfrac{1}{2} m_{\text{cylinder}} r^2 \\ &= \tfrac{1}{2}(60 \text{ lbm})(1.25 \text{ ft})^2 \\ &= 46.88 \text{ lbm-ft}^2 \end{aligned}$$

The moment, M, equals the mass moment of inertia times the angular acceleration, α.

$$\begin{aligned} M &= \frac{I_{\text{cylinder}}\alpha}{g_c} \\ rT &= \left(\frac{I_{\text{cylinder}}}{g_c}\right)\left(\frac{a}{r}\right) \end{aligned}$$

a is linear acceleration. The tension is

$$\begin{aligned} T &= \frac{I_{\text{cylinder}} a}{g_c r^2} \\ &= \frac{(46.88 \text{ lbm-ft}^2)a}{\left(32.2 \ \frac{\text{lbm-ft}}{\text{lbf-sec}^2}\right)(1.25 \text{ ft})^2} \\ &= \left(0.932 \ \frac{\text{lbf-sec}^2}{\text{ft}}\right) a \end{aligned}$$

Use a force balance around the 18 lbm weight to find another relationship between tension and acceleration.

$$\begin{aligned} \sum F &= \frac{m_{\text{weight}} a}{g_c} \\ F_{\text{weight}} - F_{\text{tension}} &= \frac{m_{\text{weight}} a}{g_c} \\ \frac{m_{\text{weight}} g}{g_c} - T &= \frac{m_{\text{weight}} a}{g_c} \\ T &= \frac{m_{\text{weight}} g}{g_c} - \left(\frac{m_{\text{weight}}}{g_c}\right) a \\ &= \frac{(18 \text{ lbm})\left(32.2 \ \frac{\text{ft}}{\text{sec}^2}\right)}{32.2 \ \frac{\text{lbm-ft}}{\text{lbf-sec}^2}} \\ &\quad - \left(\frac{18 \text{ lbm}}{32.2 \ \frac{\text{lbm-ft}}{\text{lbf-sec}^2}}\right) a \\ &= 18 \text{ lbf} - \left(0.559 \ \frac{\text{lbf-sec}^2}{\text{ft}}\right) a \end{aligned}$$

Solve the two equations simultaneously to find the acceleration.

$$\begin{aligned} T &= \left(0.932 \ \frac{\text{lbf-sec}^2}{\text{ft}}\right) a \\ T &= 18 \text{ lbf} - \left(0.559 \ \frac{\text{lbf-sec}^2}{\text{ft}}\right) a \\ 18 \text{ lbf} &= \left(1.491 \ \frac{\text{lbf-sec}^2}{\text{ft}}\right) a \\ a &= 12.07 \ \frac{\text{ft}}{\text{sec}^2} \end{aligned}$$

The tension in the wire is

$$\begin{aligned} T &= \left(0.932 \ \frac{\text{lbf-sec}^2}{\text{ft}}\right) a \\ &= \left(0.932 \ \frac{\text{lbf-sec}^2}{\text{ft}}\right)\left(12.07 \ \frac{\text{ft}}{\text{sec}^2}\right) \\ &= 11.25 \text{ lbf} \quad (11 \text{ lbf}) \end{aligned}$$

The answer is (A).

113. The average force during each of the phases of operation is

$$\begin{aligned} \bar{F}_{0-5} &= \frac{0 \text{ lbf} + 32 \text{ lbf}}{2} = 16 \text{ lbf} \\ \bar{F}_{5-9} &= \frac{32 \text{ lbf} + 20 \text{ lbf}}{2} = 26 \text{ lbf} \\ \bar{F}_{9-11} &= \frac{20 \text{ lbf} + 0 \text{ lbf}}{2} = 10 \text{ lbf} \\ \bar{F}_{11-16} &= \bar{F}_{0-5} = 16 \text{ lbf} \\ \bar{F}_{16-20} &= \bar{F}_{5-9} = 26 \text{ lbf} \end{aligned}$$

The average force during the first 20 sec of operation is

$$\begin{aligned} \bar{F}_{0-20} &= \frac{\bar{F}_{0-5}t_{0-5} + \bar{F}_{5-9}t_{5-9} + \bar{F}_{9-11}t_{9-11} + \bar{F}_{11-16}t_{11-16} + \bar{F}_{16-20}t_{16-20}}{t_{0-20}} \\ &= \frac{(16 \text{ lbf})(5 \text{ sec}) + (26 \text{ lbf})(4 \text{ sec}) + (10 \text{ lbf})(2 \text{ sec}) + (16 \text{ lbf})(5 \text{ sec}) + (26 \text{ lbf})(4 \text{ sec})}{20 \text{ sec}} \\ &= 19.4 \text{ lbf} \quad (19 \text{ lbf}) \end{aligned}$$

The answer is (D).

114. The wheel rotates as shown.

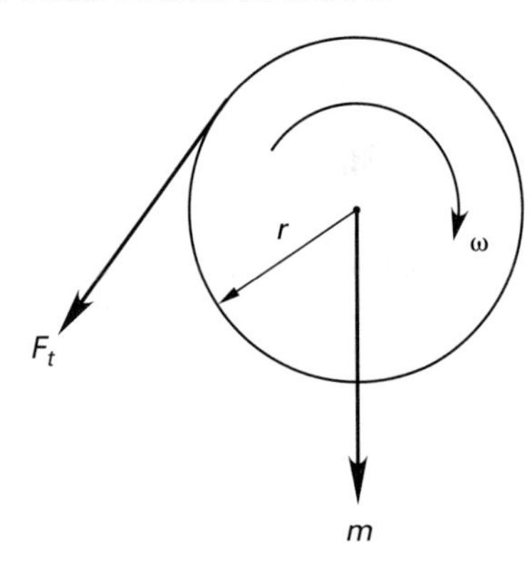

The torque, moment of inertia, angular velocity, and time are related by the equation

$$T = \left(\frac{I}{g_c}\right)\left(\frac{\Delta\omega}{\Delta t}\right)$$

The torque is also equal to the tangential force times the radius.

$$T = F_t r$$

The wheel is a solid cylinder, so its moment of inertia is

$$I = \frac{mr^2}{2}$$

Substituting into the first equation and solving for the tangential force gives

$$F_t r = \left(\frac{mr^2}{2g_c}\right)\left(\frac{\Delta\omega}{\Delta t}\right)$$

$$F_t = \frac{mr\Delta\omega}{2g_c\Delta t}$$

Reducing the speed of 54 rad/sec by a third means a change of 18 rad/sec. The tangential force needed to alter the angular velocity by 18 rad/sec in 30 sec is

$$F_t = \frac{(90 \text{ lbm})(5 \text{ ft})\left(18 \ \frac{\text{rad}}{\text{sec}}\right)}{(2)\left(32.2 \ \frac{\text{lbm-ft}}{\text{lbf-sec}^2}\right)(30 \text{ sec})}$$

$$= 4.19 \text{ lbf} \quad (4 \text{ lbf})$$

The answer is (A).

115. The effective load is greater than the actual load due to the speed of application (impact). Calculate the effective load from the impact factor equation. Set the height of impact to zero.

$$F_e = P\left(1 + \sqrt{1 + \frac{2h}{\delta_{st}}}\right)$$

$$= (1000 \text{ lbf})\left(1 + \sqrt{1 + \frac{(2)(0 \text{ ft})}{\delta_{st}}}\right)$$

$$= 2000 \text{ lbf}$$

Calculate the weld parameters treating the weld as a line.

$$A_w = 2L_w = (2)(10 \text{ in}) = 20 \text{ in}$$

$$S_w = \frac{L_w^2}{3} = \frac{(10 \text{ in})^2}{3} = 33.3 \text{ in}^2$$

$$M = (2000 \text{ lbf})(2 \text{ in}) = 4000 \text{ in-lbf}$$

Calculate forces on the weld. The direct shear is

$$V = P = 2000 \text{ lbf}$$

$$f_s = \frac{P}{A_w} = \frac{2000 \text{ lbf}}{20 \text{ in}} = 100 \text{ lbf/in}$$

The shear due to bending is

$$f_b = \frac{M}{S_w} = \frac{4000 \text{ in-lbf}}{33.3 \text{ in}^2} = 120 \text{ lbf/in}$$

Calculate the resolved shear force.

$$f = \sqrt{f_s^2 + f_b^2}$$

$$= \sqrt{\left(100 \ \frac{\text{lbf}}{\text{in}}\right)^2 + \left(120 \ \frac{\text{lbf}}{\text{in}}\right)^2}$$

$$= 156.2 \text{ lbf/in}$$

Calculate the weld force capacity. The allowable force per inch of weld leg is 5000 lbf/in. There are two welds, each with 1/8 in leg. (The weld throat is smaller than the weld leg, but the weld resistance is given per inch of weld leg, not weld throat.)

$$f_{\text{allowable}} = \left(\frac{5000 \ \frac{\text{lbf}}{\text{in}}}{\text{in}}\right)(2 \text{ legs})\left(\frac{1}{8} \ \frac{\text{in}}{\text{leg}}\right)$$

$$= 1250 \text{ lbf/in}$$

Calculate the factor of safety.

$$\text{FS} = \frac{f_{\text{allowable}}}{f} = \frac{1250 \ \frac{\text{lbf}}{\text{in}}}{156.2 \ \frac{\text{lbf}}{\text{in}}}$$

$$= 8.0$$

The answer is (C).

116. Definitions from the American Welding Society are:

Soldering – A joining process wherein coalescence between metal parts is produced by heating to suitable temperatures generally below 800°F and by using nonferrous filler metals having melting temperatures below those of the base metals. The solder is usually distributed between the properly fitted surfaces of the joint by capillary attraction.

Brazing – A group of welding processes wherein coalescence is produced by heating to suitable temperatures above 800°F and by using a nonferrous filler metal having a melting point below those of the base metals. The filler metal is distributed between closely fitted surfaces of the joint by capillary action.

Welding – A localized coalescence of metals wherein coalescence is produced by heating to suitable temperatures, with or without the application of pressure, and with or without the use of filler metal. The filler metal either has a melting point approximately the same as the base metals, or has a melting point below that of the base metals but above 800°F.

Forge Welding – A group of welding processes wherein coalescence is produced by heating in a forge or other furnace and by applying pressure or blows. An example is the hammer welding process previously used in railroad and blacksmith shops.

The answer is (A).

117. Consult tables of fits that are available in most design references. The following table is from ANSI standard B4.1-1967, R1987.

class RC 7

nominal range over–to	clearance (10^{-3} in)	standard tolerance limits (10^{-3} in) hole H9	shaft d8
0.40–0.71	2.0	+1.6	−2.0
	4.6	0	−3.0

The answer is (A).

118. The logarithmic decrement is the natural logarithm of the ratio of two successive amplitudes. It is related to the damping ratio by the equation

$$\ln\frac{x_n}{x_{n+1}} = \frac{2\pi\zeta}{\sqrt{1-\zeta^2}}$$

Squaring both sides and solving for the damping ratio,

$$\left(\ln\frac{x_n}{x_{n+1}}\right)^2 = \frac{4\pi^2\zeta^2}{1-\zeta^2}$$

$$4\pi^2\zeta^2 = \left(\ln\frac{x_n}{x_{n+1}}\right)^2 - \zeta^2\left(\ln\frac{x_n}{x_{n+1}}\right)^2$$

$$\zeta^2\left(4\pi^2 + \left(\ln\frac{x_n}{x_{n+1}}\right)^2\right) = \left(\ln\frac{x_n}{x_{n+1}}\right)^2$$

$$\zeta = \sqrt{\frac{\left(\ln\frac{x_n}{x_{n+1}}\right)^2}{4\pi^2 + \left(\ln\frac{x_n}{x_{n+1}}\right)^2}} = \sqrt{\frac{\left(\ln\frac{0.569}{0.462}\right)^2}{4\pi^2 + \left(\ln\frac{0.569}{0.462}\right)^2}} = 0.0331 \quad (0.033)$$

The answer is (B).

119. Consider the overall geometry of the truss. The tension in the wire is equal to the weight.

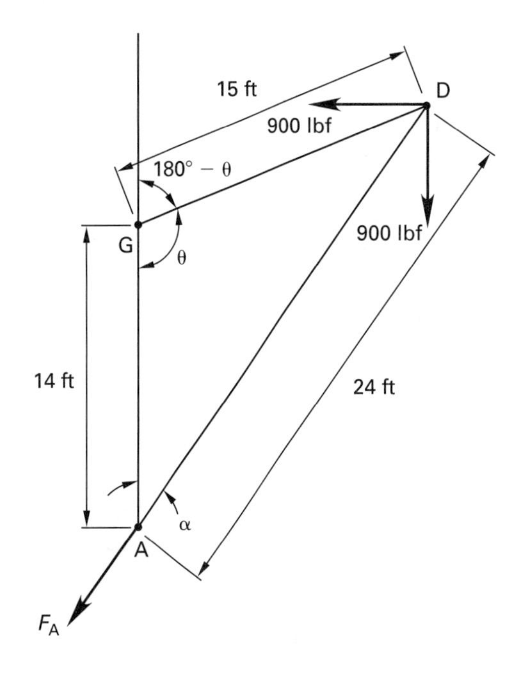

Calculate angles θ and α with the law of cosines.

$$a^2 = b^2 + c^2 - 2bc\cos A$$

$$\cos A = \frac{b^2+c^2-a^2}{2bc}$$

$$\cos\theta = \frac{(\mathrm{AG})^2 + (\mathrm{DG})^2 - (\mathrm{AD})^2}{2(\mathrm{AG})(\mathrm{DG})} = \frac{(14\text{ ft})^2 + (15\text{ ft})^2 - (24\text{ ft})^2}{(2)(14\text{ ft})(15\text{ ft})} = -0.369$$

$$\theta = 112°$$

$$\cos\alpha = \frac{(\mathrm{AG})^2 + (\mathrm{AD})^2 - (\mathrm{DG})^2}{2(\mathrm{AG})(\mathrm{AD})} = \frac{(14\text{ ft})^2 + (24\text{ ft})^2 - (15\text{ ft})^2}{(2)(14\text{ ft})(24\text{ ft})} = 0.814$$

$$\alpha = 35.5°$$

For static equilibrium, the sum of the moments about point G is zero. The pin connection of the two-force member AB can support only a force in the direction of the member AB.

$$\begin{aligned}\sum M_G &= 0 \text{ ft-lbf} \\ &= (\text{AG})F_{\text{A}} \sin\alpha + (\text{DG})F_w \sin(180° - \theta) \\ &\quad - (\text{DG})F_w \cos(180° - \theta) \\ F_{\text{A}} &= \frac{(\text{DG})F_w\left(\cos(180° - \theta) - \sin(180° - \theta)\right)}{(\text{AG})\sin\alpha} \\ &= \frac{(15 \text{ ft})(900 \text{ lbf})\begin{pmatrix}\cos(180° - 112°) \\ - \sin(180° - 112°)\end{pmatrix}}{(14 \text{ ft})\sin 35.5°} \\ &= -917.58 \text{ lbf}\end{aligned}$$

All members of a truss are two-force members, so connecting joints do not transmit moments, and each force acts in the direction of the member. A force balance around point B gives

$$\begin{aligned}F_{\text{BG}} &= F_{\text{BF}} = 0 \text{ lbf} \\ F_{\text{A}} &= F_{\text{BC}} = -917.58 \text{ lbf} \quad (920 \text{ lbf, compression})\end{aligned}$$

The answer is (C).

120. The nominal stack height is the sum of the nominal part heights.

$$\begin{aligned}N &= N_1 + N_2 + N_3 \\ &= 0.8 \text{ in} + 1.0 \text{ in} + 1.2 \text{ in} \\ &= 3.0 \text{ in}\end{aligned}$$

The stack tolerance is a root sum square combination of parts tolerances.

$$\begin{aligned}T &= \sqrt{T_1^2 + T_2^2 + T_3^2} \\ &= \sqrt{(0.1 \text{ in})^2 + (0.2 \text{ in})^2 + (0.3 \text{ in})^2} \\ &= 0.37 \text{ in} \quad (0.4 \text{ in})\end{aligned}$$

The toleranced stack height combines the stack nominal height and tolerance. Given that all part tolerances are normally distributed and $\pm 3\sigma$ from their means, this value represents 99.7% of stack assemblies.

$$N + T = 3.0 \text{ in} \pm 0.4 \text{ in}$$

The answer is (C).

Solutions
Thermal and Fluids Systems

121. The valve flow coefficient, C_v, relates flow rate to the pressure drop by the following equation.

$$Q_{gpm} = C_v \sqrt{\frac{\Delta p_{psi}}{SG}}$$

The value flow coefficient is related to the loss coefficient by the equation

$$C_v = \frac{(29.9)(d_{in})^2}{\sqrt{K}}$$

Combining the two equations and solving for the valve diameter,

$$Q_{gpm} = \frac{(29.9)(d_{in})^2}{\sqrt{K}} \sqrt{\frac{\Delta p_{psi}}{SG}}$$

$$d_{in} = \sqrt{\frac{Q_{gpm}\sqrt{K}}{29.9\sqrt{\frac{\Delta p_{psi}}{SG}}}}$$

For a fully open gate value, the loss coefficient is $K = 0.19$. The minimum needed diameter, then, is

$$d_{in} = \sqrt{\frac{\left(60\ \frac{\text{gal}}{\text{min}}\right)\sqrt{0.19}}{29.9\sqrt{\frac{5\ \frac{\text{lbf}}{\text{in}^2}}{1}}}} = 0.63\ \text{in}$$

Of the available choices, the smallest adequate choice is $d_{in} = 3/4$ in.

The answer is (B).

122. In order to lift a ball, the drag force created by the air on the ball needs to be greater than the ball's weight. The air velocity that makes the drag force equal to the weight will be the minimum velocity.

$$F_D = W$$

$$\frac{\rho C_D A v^2}{2g_c} = W$$

$$\left(\frac{1}{2}\right)\left(\frac{0.075\ \frac{\text{lbm}}{\text{ft}^3}}{32.2\ \frac{\text{ft-lbm}}{\text{lbf-sec}^2}}\right) \times C_D\left(\frac{\pi}{4}\right)\left((1.5\ \text{in})\left(\frac{1\ \text{ft}}{12\ \text{in}}\right)\right)^2 v^2 = 0.00516\ \text{lbf}$$

$$\left(1.43\times 10^{-5}\ \frac{\text{lbf-sec}^2}{\text{ft}^2}\right) C_D v^2 = 0.00516\ \text{lbf}$$

The Reynolds number is needed to determine the drag coefficient, C_D, from a drag coefficient graph.

$$\text{Re} = \frac{vD}{\nu}$$

But to find the Reynolds number, the velocity, another unknown variable, is needed. Therefore, the solution is iterative and requires an assumed drag coefficient. On a graph of the Reynolds number for a sphere, the drag coefficient varies between 0.4 and 0.45 for Reynolds numbers between 1×10^3 and 1×10^5. Use an initial drag coefficient of 0.4. With this value, the minimum velocity becomes

$$\left(1.43\times 10^{-5}\ \frac{\text{lbf-sec}^2}{\text{ft}^2}\right)(0.4)v^2 = 0.00516\ \text{lbf}$$

$$v = 30.0\ \text{ft/sec}$$

Check the Reynolds number to verify the assumed value of the drag coefficient.

$$\text{Re} = \frac{vD}{\nu} = \frac{\left(30.0\ \frac{\text{ft}}{\text{sec}}\right)(1.5\ \text{in})\left(\frac{1\ \text{ft}}{12\ \text{in}}\right)}{1.58\times 10^{-4}\ \frac{\text{ft}^2}{\text{sec}}} = 2.4\times 10^4$$

With this Reynolds number, the drag coefficient for a sphere is approximately 0.42, close to the initial assumption.

The flow rate through the container is

$$\dot{V} = vA = \left(30.0\ \frac{\text{ft}}{\text{sec}}\right)(2\ \text{ft})(2\ \text{ft})\left(60\ \frac{\text{sec}}{\text{min}}\right) = 7200\ \text{ft}^3/\text{min}$$

The answer is (B).

123. The power generated by the turbine is

$$P = \dot{V}\gamma h_t$$

The velocity head of the jet is

$$h_t = \frac{v_j^2}{2g}$$

The turbine power becomes

$$P = \dot{V}\gamma \frac{v_j^2}{2g}$$

Find the flow rate.

$$\begin{aligned} \dot{V} &= v_j A_j \\ &= v_j\left(\frac{\pi}{4}\right)(2\ \text{ft})^2 \\ &= (\pi\ \text{ft}^2)v_j \end{aligned}$$

Substitute known values.

$$(122 \times 10^6\ \text{W})\left(\frac{1\ \text{hp}}{745.7\ \text{W}}\right)\left(550\ \frac{\text{ft-lbf}}{\text{hp-sec}}\right) = (\pi\ \text{ft}^2)v_j\left(62.4\ \frac{\text{lbf}}{\text{ft}^3}\right)\left(\frac{v_j^2}{(2)\left(32.2\ \frac{\text{ft}}{\text{sec}^2}\right)}\right)$$

$$v_j = 309\ \text{ft/sec}$$

For ideal conditions, the tangential speed of the vanes is half the speed of the jet. Find the angular velocity of the turbine wheel and convert it to rpm.

$$\begin{aligned} v_{\text{vane}} &= r\omega \\ 0.5v_j &= r\omega \\ (0.5)\left(309\ \frac{\text{ft}}{\text{sec}}\right) &= (10\ \text{ft})\omega \\ \omega &= 15.5\ \text{rad/sec} \\ &= \left(\frac{1\ \text{rev}}{2\pi\ \text{rad}}\right)\left(15.5\ \frac{\text{rad}}{\text{sec}}\right)\left(60\ \frac{\text{sec}}{\text{min}}\right) \\ &= 148\ \text{rpm}\quad (150\ \text{rpm}) \end{aligned}$$

The answer is (B).

124. The easiest way to locate the points of maximum and minimum pressure is to draw the hydraulic grade line (HGL).

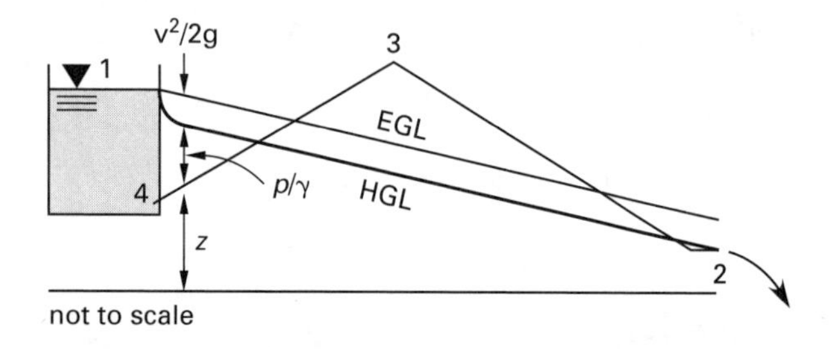

The height p/γ represents the distance to which the liquid in a manometer would rise if a manometer were inserted into the pipe at that point. The height of the p/γ line along the length of the system is known as the hydraulic grade line, HGL. The total head in the system is greater than $p/\gamma + z$ by an amount $v^2/2g$, so the energy gradeline, EGL, is above the HGL by a distance of $v^2/2g$. If the pipe rises above the HGL, then p/γ is negative. The advantage of drawing the HGL is that it makes apparent where the minimum and maximum pressures in the pipe occur.

The maximum pressure in the system occurs at point 4, before the water enters the pipe. Note the initial rapid drop in p/γ right after the water enters the pipe. This happens because some of the pressure head has been converted into velocity head. The maximum pressure in the system is simply the static pressure at point 4.

$$\begin{aligned} p_4 = \gamma h &= \left(62.4\ \frac{\text{lbf}}{\text{ft}^3}\right)(330\ \text{ft} - 315\ \text{ft})\left(\frac{1\ \text{ft}}{12\ \text{in}}\right)^2 \\ &= 6.5\ \text{lbf/in}^2 \end{aligned}$$

The minimum pressure will occur at point 3 since it is the furthest point above the HGL and will, therefore, be the most negative in the system. To compute the pressure at this point, however, the velocity of the water in the pipe must be calculated. The velocity can be found by writing the Bernoulli equation between points 1 and 2.

$$\frac{p_1}{\gamma} + \frac{v_1^2}{2g} + z_1 + h_p = \frac{p_2}{\gamma} + \frac{v_2^2}{2g} + z_2 + h_L$$

$$p_1 = p_2 = 0$$
$$v_1 = 0$$
$$h_p = 0$$

Substituting these values and the equation for h_L yields

$$330\ \text{ft} = \frac{v^2}{2g} + 100\ \text{ft} + 0.015\left(\frac{L}{D}\right)\left(\frac{v^2}{2g}\right)$$

$$230\ \text{ft} = \frac{v^2}{(2)\left(32.2\ \frac{\text{ft}}{\text{sec}^2}\right)} + (0.015)\left(\frac{1650\ \text{ft}}{2\ \text{ft}}\right) \times \left(\frac{v^2}{(2)\left(32.2\ \frac{\text{ft}}{\text{sec}^2}\right)}\right)$$

$$v = 33.3\ \text{ft/sec}$$

The minimum pressure can be found by writing the Bernoulli equation between points 1 and 3.

$$\frac{p_1}{\gamma} + \frac{v_1^2}{2g} + z_1 + h_p = \frac{p_3}{\gamma} + \frac{v_3^2}{2g} + z_3 + h_L$$

$$p_1 = 0$$
$$v_1 = 0$$
$$h_p = 0$$

Substituting these values and the equation for h_L yields

$$330 \text{ ft} = \frac{p_3}{62.4 \frac{\text{lbf}}{\text{ft}^3}} + \frac{\left(33.3 \frac{\text{ft}}{\text{sec}}\right)^2}{(2)\left(32.2 \frac{\text{ft}}{\text{sec}^2}\right)} + 333 \text{ ft}$$

$$+ (0.015)\left(\frac{350 \text{ ft}}{2 \text{ ft}}\right)\left(\frac{\left(33.3 \frac{\text{ft}}{\text{sec}}\right)^2}{(2)\left(32.2 \frac{\text{ft}}{\text{sec}^2}\right)}\right)$$

$$p_3 = \left(-4082 \frac{\text{lbf}}{\text{ft}^2}\right)\left(\frac{1 \text{ ft}}{12 \text{ in}}\right)^2$$

$$= -28.3 \text{ lbf/in}^2 \quad (-28 \text{ lbf/in}^2)$$

The answer is (A).

125. The power that must be supplied to the water is

$$P = \dot{V}\gamma h_p$$

Use the efficiencies of the motor and pump, and substitute known values.

$$(0.9)(0.8)P = \left(1500 \frac{\text{gal}}{\text{min}}\right)\left(\frac{1 \text{ ft}^3}{7.48 \text{ gal}}\right) \times \left(\frac{1 \text{ min}}{60 \text{ sec}}\right)\left(62.4 \frac{\text{lbf}}{\text{ft}^3}\right) h_p$$

$$0.72P = \left(208.6 \frac{\text{lbf}}{\text{sec}}\right) h_p$$

The pump head can be found using the Bernoulli equation.

$$\frac{p_1}{\gamma} + \frac{\text{v}_1^2}{2g} + z_1 + h_p = \frac{p_2}{\gamma} + \frac{\text{v}_2^2}{2g} + z_2 + h_L$$

Define points 1 and 2 to be the surfaces of the lower and upper tanks, respectively.

$$p_1 = p_2 = 0$$
$$\text{v}_1 = \text{v}_2 = 0$$

v can be found using the flow rate equation.

$$\dot{V} = \text{v}A$$

$$\left(1500 \frac{\text{gal}}{\text{min}}\right)\left(\frac{1 \text{ ft}^3}{7.48 \text{ gal}}\right)\left(\frac{1 \text{ min}}{60 \text{ sec}}\right) = \text{v}\left(\frac{\pi}{4}\right)\left((8 \text{ in})\left(\frac{1 \text{ ft}}{12 \text{ in}}\right)\right)^2$$

$$\text{v} = 9.57 \text{ ft/sec}$$

Substitute known values and the equation for the head loss in the pipe.

$$125 \text{ ft} + h_p = 200 \text{ ft} + 0.02\left(\frac{L}{D}\right)\left(\frac{\text{v}^2}{2g}\right)$$

$$h_p = 75 \text{ ft} + (0.02)\left(\frac{4000 \text{ ft}}{(8 \text{ in})\left(\frac{1 \text{ ft}}{12 \text{ in}}\right)}\right) \times \left(\frac{\left(9.57 \frac{\text{ft}}{\text{sec}}\right)^2}{(2)\left(32.2 \frac{\text{ft}}{\text{sec}^2}\right)}\right)$$

$$= 245.7 \text{ ft}$$

The power supplied to the motor is

$$0.72P = \left(208.6 \frac{\text{lbf}}{\text{sec}}\right)(245.7 \text{ ft}) \times \left(\frac{1 \text{ hp}}{550 \frac{\text{ft-lbf}}{\text{sec}}}\right)\left(745.7 \frac{\text{W}}{\text{hp}}\right)$$

$$P = 96{,}500 \text{ W} \quad (97 \text{ kW})$$

The answer is (C).

126. A pump curve represents the head a pump can supply for a given flow rate. To maintain a prescribed flow rate in a system, a certain amount of head must be supplied to the system. As the flow rate in the system increases, the amount of head that must be supplied also increases. By plotting the head needed by the system versus the flow rate (the system curve) on top of the pump curve, the point where the head produced by the pump is just the amount needed to overcome the head loss in the pipe can be found. The point where this occurs is where the two curves intersect and is known as the operating point.

An equation for the system curve can be obtained using the Bernoulli equation.

$$\frac{p_1}{\gamma} + \frac{\text{v}_1^2}{2g} + z_1 + h_p = \frac{p_2}{\gamma} + \frac{\text{v}_2^2}{2g} + z_2 + h_L$$

Define points 1 and 2 to be the surface of the lower and upper tanks, respectively.

$$p_1 = p_2 = 0$$
$$\text{v}_1 = \text{v}_2 = 0$$

Substitute known values and the equation for the pipe head loss into the Bernoulli equation.

$$50 \text{ ft} + h_p = 150 \text{ ft} + \frac{fL\text{v}^2}{2Dg}$$

The equivalent lengths of the fittings and valves are negligible compared to the total pipe length.

$$h_p = 100\ \text{ft} + \frac{(0.015)(3500\ \text{ft})\text{v}^2}{(2)(16\ \text{in})\left(\frac{1\ \text{ft}}{12\ \text{in}}\right)\left(32.2\ \frac{\text{ft}}{\text{sec}^2}\right)}$$

Express v in terms of $\dot{V}$.

$$\begin{aligned} \dot{V} &= \text{v}A \\ &= \text{v}\left(\frac{\pi}{4}\right)\left((16\ \text{in})\left(\frac{1\ \text{ft}}{12\ \text{in}}\right)\right)^2 \\ \text{v} &= (0.716\ \text{ft}^{-2})\dot{V} \end{aligned}$$

Substitute into the Bernoulli equation.

$$\begin{aligned} h_p &= 100\ \text{ft} + \left(\frac{(0.015)(3500\ \text{ft})\left((0.716\ \text{ft}^{-2})\dot{V}\right)^2}{(2)(16\ \text{in})\left(\frac{1\ \text{ft}}{12\ \text{in}}\right)\left(32.2\ \frac{\text{ft}}{\text{sec}^2}\right)}\right) \\ &= 100\ \text{ft} + \left(0.313\ \frac{\text{sec}^2}{\text{ft}^5}\right)\dot{V}^2 \end{aligned}$$

Plot the curve.

Q (gpm)	$\dot{V}$ (ft³/sec)	h_p (ft)
1200	2.674	102.2
2400	5.348	109.0
3600	8.022	120.1
4800	10.70	135.8
6000	13.37	156.0

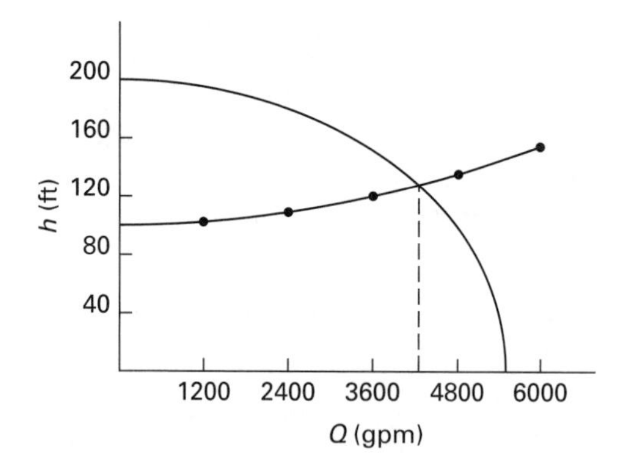

The discharge rate is approximately 4200 gpm.

$$Q = 4200\ \text{gpm}$$

The answer is (C).

127. Enthalpy is

$$h = u + \frac{pv}{J} = u + \frac{p}{J\rho}$$

The change in enthalpy from A to B is

$$\begin{aligned} \Delta h &= h_\text{B} - h_\text{A} \\ &= u_\text{B} - u_\text{A} + \frac{\frac{p_\text{B}}{\rho_\text{B}} - \frac{p_\text{A}}{\rho_\text{A}}}{J} \\ &= 53.1\ \frac{\text{Btu}}{\text{lbm}} - 119.1\ \frac{\text{Btu}}{\text{lbm}} \\ &\quad + \frac{\left(\frac{5.85\ \frac{\text{lbf}}{\text{in}^2}}{0.0508\ \frac{\text{lbm}}{\text{ft}^3}} - \frac{98.4\ \frac{\text{lbf}}{\text{in}^2}}{0.381\ \frac{\text{lbm}}{\text{ft}^3}}\right)\left(144\ \frac{\text{in}^2}{\text{ft}^2}\right)}{778\ \frac{\text{ft-lbf}}{\text{Btu}}} \\ &= -92.5\ \text{Btu/lbm} \quad (-93\ \text{Btu/lbm}) \end{aligned}$$

The answer is (A).

128. Solve for the drag coefficient.

$$\begin{aligned} C_D &= \frac{4Dg(\rho_\text{sphere} - \rho_\text{fluid})}{3\text{v}^2\rho_\text{fluid}} \\ &= \frac{(4)(3\ \text{in})\left(386.4\ \frac{\text{in}}{\text{sec}^2}\right)\left(488\ \frac{\text{lbm}}{\text{ft}^3} - 87\ \frac{\text{lbm}}{\text{ft}^3}\right)}{(3)\left(122\ \frac{\text{in}}{\text{sec}}\right)^2\left(87\ \frac{\text{lbm}}{\text{ft}^3}\right)} \\ &= 0.479 \quad (0.48) \end{aligned}$$

The answer is (D).

129. Use the Bernoulli equation.

$$\frac{p_1}{\gamma} + \frac{\text{v}_1^2}{2g} + z_1 + h_p = \frac{p_2}{\gamma} + \frac{\text{v}_2^2}{2g} + z_2 + h_L$$

Define point 1 as the surface of the tank and point 2 as the top of the fountain. At both points, the velocity of the stream is zero.

$$\begin{aligned} p_1 &= p_2 = 0 \\ \text{v}_1 &= \text{v}_2 = 0 \end{aligned}$$

The Bernoulli equation simplifies to

$$z_1 = 4\ \text{ft} + h_L$$

The head loss is

$$h_L = \frac{K_\text{entering}\text{v}^2}{2g} + \frac{fL\text{v}^2}{2gD}$$

The velocity of the water at the tank exit, and therefore in the pipe, is

$$\text{v} = \sqrt{2gh} = \sqrt{(2)\left(32.2\ \frac{\text{ft}}{\text{sec}^2}\right)(4\ \text{ft})} = 16.0\ \text{ft/sec}$$

For a sharp edge exit,

$$K_{\text{entering}} = 0.5$$

The effective length is

$$L = 40 \text{ ft} + (5)(4.6 \text{ ft}) + 2.5 \text{ ft} = 65.5 \text{ ft}$$

Calculate the Reynolds number.

$$\text{Re}_D = \frac{\text{v}D}{\nu} = \frac{\left(16 \ \frac{\text{ft}}{\text{sec}}\right)(0.3355 \text{ ft})}{1.4 \times 10^{-5} \ \frac{\text{ft}^2}{\text{sec}}} = 3.83 \times 10^5$$

$$\frac{\varepsilon}{D} = \frac{0.0002 \text{ ft}}{0.3355 \text{ ft}} = 0.0006$$

Using this Reynolds number and this relative roughness, the friction factor is found from the Moody chart to be $f = 0.019$. The head loss is

$$\begin{aligned} h_L &= \frac{(0.5)\left(16 \ \frac{\text{ft}}{\text{sec}}\right)^2}{(2)\left(32.2 \ \frac{\text{ft}}{\text{sec}^2}\right)} + \frac{(0.019)(65.5 \text{ ft})\left(16 \ \frac{\text{ft}}{\text{sec}}\right)^2}{(2)\left(32.2 \ \frac{\text{ft}}{\text{sec}^2}\right)(0.3355 \text{ ft})} \\ &= 16.7 \text{ ft} \end{aligned}$$

The required depth in the tank is

$$\begin{aligned} z_1 &= 4 \text{ ft} + 16.7 \text{ ft} \\ &= 20.7 \text{ ft} \quad (21 \text{ ft}) \end{aligned}$$

The answer is (B).

130. The volume of a fluid flowing through a pipe is the product of the velocity and the free cross-sectional area of the pipe.

$$\dot{V} = \text{v}A_{\text{pipe}}$$

The minimum cross-sectional area required to keep the water velocity under 7 ft/sec is

$$\begin{aligned} A_{\text{pipe}} &= \frac{\dot{V}}{\text{v}} \\ &= \frac{\left(100 \ \frac{\text{gal}}{\text{min}}\right)\left(0.13368 \ \frac{\text{ft}^3}{\text{gal}}\right)}{\left(7.0 \ \frac{\text{ft}}{\text{sec}}\right)\left(60 \ \frac{\text{sec}}{\text{min}}\right)} \\ &= 0.0318 \text{ ft}^2 \end{aligned}$$

The smallest schedule-40 pipe that provides the required area is $2^1/_2$ in nominal diameter, which has a cross-sectional area of 0.03325 ft^2.

The answer is (B).

131. Find the stroke length in feet.

$$L = \frac{8 \text{ in}}{12 \ \frac{\text{in}}{\text{ft}}} = 0.667 \text{ ft}$$

Find the bore area in square inches.

$$\begin{aligned} A &= \frac{\pi}{4}D^2 \\ &= \left(\frac{\pi}{4}\right)(8 \text{ in})^2(2) \\ &= 50.27 \text{ in}^2 \end{aligned}$$

Find the number of engine power strokes per minute. The Otto engine operates on a four-stroke cycle.

$$\begin{aligned} N &= \frac{(\text{no. of strokes per revolution})n \times (\text{no. of cylinders})}{\text{no. of strokes per power stroke}} \\ &= \frac{\left(2 \ \frac{\text{strokes}}{\text{rev}}\right)\left(400 \ \frac{\text{rev}}{\text{min}}\right)(2)}{\frac{4 \text{ strokes}}{\text{power stroke}}} \\ &= 400 \text{ power strokes/min} \end{aligned}$$

Use the PLAN formula to solve for the brake mean effective pressure.

$$\begin{aligned} \text{BHP} &= (\text{BMEP})(LAN) \\ \text{BMEP} &= \frac{\text{BHP}}{LAN} \\ &= \frac{(33 \text{ hp})\left(33{,}000 \ \frac{\text{ft-lbf}}{\text{hp-min}}\right)}{(0.667 \text{ ft})(50.27 \text{ in}^2)\left(400 \ \frac{\text{power strokes}}{\text{min}}\right)} \\ &= 81.2 \text{ lbf/in}^2 \quad (81 \text{ psig}) \end{aligned}$$

The answer is (D).

132. Find the flow through the valve. The minimum velocity of the cylinder is

$$\begin{aligned} \text{v} &= \frac{L}{t} = \frac{24 \text{ in}}{5 \text{ sec}} \\ &= 4.8 \text{ in/sec} \end{aligned}$$

The effective area of the cylinder is

$$\begin{aligned} A &= \frac{\pi}{4}\left(d_{\text{bore}}^2 - d_{\text{rod}}^2\right) \\ &= \frac{\pi}{4}\left((3 \text{ in})^2 - (1 \text{ in})^2\right) \\ &= 6.28 \text{ in}^2 \end{aligned}$$

The flow rate is

$$Q = \mathrm{v}A = \frac{\left(4.8\ \frac{\text{in}}{\text{sec}}\right)(6.28\ \text{in}^2)\left(60\ \frac{\text{sec}}{\text{min}}\right)}{231\ \frac{\text{in}^3}{\text{gal}}}$$

$$= 7.830\ \text{gpm}$$

Valve flow is related to the pressure drop by the valve flow coefficient.

$$Q_{\text{gpm}} = C_{\text{v}}\sqrt{\frac{\Delta p_{\text{psi}}}{\text{SG}}}$$

$$\Delta p_{\text{psi}} = \left(\frac{Q_{\text{gpm}}}{C_{\text{v}}}\right)^2 \text{SG}$$

$$= \left(\frac{7.830\ \frac{\text{gal}}{\text{min}}}{2.9}\right)^2 (0.85)$$

$$= 6.20\ \text{psi}$$

The answer is (A).

133. The work for a heat pump can be found from

$$\text{COP} = \frac{\dot{Q}_H}{P}$$

COP is the coefficient of performance. The minimum cost will occur if the heat pump operates at ideal efficiency. For a Carnot heat pump, the COP is

$$\text{COP} = \frac{1}{1 - \frac{T_L}{T_H}}$$

T is in degrees Rankine. The heat pump power is

$$P = \left(1 - \frac{T_L}{T_H}\right)\dot{Q}_H$$

$$= \left(1 - \frac{30°\text{F} + 460°}{78°\text{F} + 460°}\right)\left(900{,}000\ \frac{\text{Btu}}{\text{day}}\right)$$

$$= 80{,}300\ \text{Btu/day}$$

$$\text{Cost per month} = \left(80{,}300\ \frac{\text{Btu}}{\text{day}}\right)\left(\frac{1\ \text{kW-hr}}{3412\ \text{Btu}}\right) \times \left(0.037\ \frac{\$}{\text{kW-hr}}\right)\left(30\ \frac{\text{days}}{\text{month}}\right)$$

$$= \$26.12 \quad (\$26)$$

The answer is (B).

134. Assume the thermal conductivity of the uranium fuel plate is constant. For a flat plate with an internal heat source, constant thermal conductivity, and uniform heat generation, the heat source generation rate, G, is

$$G = -k\frac{d^2T}{dx^2}$$

The thermal conductivity of uranium is 20 Btu/hr-ft-°F.

$$9 \times 10^7\ \frac{\text{Btu}}{\text{hr-ft}^3} = -\left(20\ \frac{\text{Btu}}{\text{hr-ft}^3}\right)\frac{d^2T}{dx^2}$$

$$\frac{d^2T}{dx^2} = -4.5 \times 10^6\ °\text{F/ft}^2$$

Integrating twice,

$$\frac{dT}{dx} = \left(-4.5 \times 10^6\ \frac{°\text{F}}{\text{ft}^2}\right)x + C_1$$

$$T = \left(-2.25 \times 10^6\ \frac{°\text{F}}{\text{ft}^2}\right)x^2 + C_1x + C_2$$

Use the boundary conditions to find the constants of integration. At $x = 0$, $T = 500°\text{F}$, so $C_2 = 500°\text{F}$. At $x = 0.25$ in, $T = 750°\text{F}$.

$$750°\text{F} = \left(-2.25 \times 10^6\ \frac{°\text{F}}{\text{ft}^2}\right)\left((0.25\ \text{in})\left(\frac{1\ \text{ft}}{12\ \text{in}}\right)\right)^2 + C_1(0.25\ \text{in})\left(\frac{1\ \text{ft}}{12\ \text{in}}\right) + 500°\text{F}$$

$$C_1 = 58{,}875\ °\text{F/ft}$$

The equation for T is

$$T = \left(-2.25 \times 10^6\ \frac{°\text{F}}{\text{ft}^2}\right)x^2 + \left(58{,}875\ \frac{°\text{F}}{\text{ft}}\right)x + 500°\text{F}$$

The temperature is maximum when $dT/dx = 0$.

$$0 = -\left(4.5 \times 10^6\ \frac{°\text{F}}{\text{ft}^2}\right)x + 58{,}875\ \frac{°\text{F}}{\text{ft}}$$

$$x = 0.0131\ \text{ft}$$

$$T_{\text{max}} = \left(-2.25 \times 10^6\ \frac{°\text{F}}{\text{ft}^2}\right)(0.0131\ \text{ft})^2 + \left(58{,}875\ \frac{°\text{F}}{\text{ft}}\right)(0.0131\ \text{ft}) + 500°\text{F}$$

$$= 885°\text{F} \quad (890°\text{F})$$

The answer is (B).

135. The changes in temperature from the beginning state to the end state can be graphed as shown.

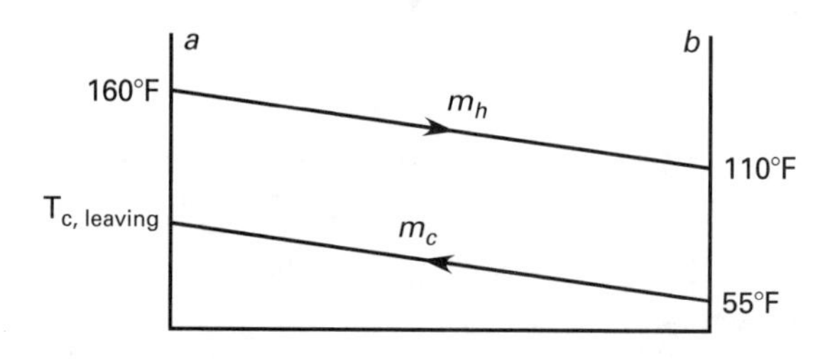

The outer surface area can be found from

$$q = UA(\text{LMTD})$$

The LMTD is the log mean temperature difference. Since $m_h c_{ph} = m_c c_{pc}$, the LMTD is 160°F − 110°F = 50°F. This is true only for counterflow heat exchangers. Thus, $T_{c,\text{leaving}}$ does not need to be calculated. If $m_h c_{ph} \neq m_c c_{pc}$, then it would be necessary to compute $T_{c,\text{leaving}}$ from a heat balance of the two fluids.

Find the surface area.

$$\begin{aligned} q &= UA(\text{LMTD}) \\ 2.03\times 10^6\ \frac{\text{Btu}}{\text{hr}} &= \left(75\ \frac{\text{Btu}}{\text{hr-ft}^2\text{-°F}}\right) A\,(50\text{°F}) \\ A &= 541\ \text{ft}^2 \end{aligned}$$

The number of tubes is

$$\begin{aligned} N &= \frac{A}{A_\text{tube}} = \frac{541\ \text{ft}^2}{\pi DL} \\ &= \frac{541\ \text{ft}^2}{\pi\,(1\ \text{in})\left(\frac{1\ \text{ft}}{12\ \text{in}}\right)(30\ \text{ft})} \\ &= 68.8\ \text{tubes} \quad (69\ \text{tubes}) \end{aligned}$$

The answer is (B).

136. The efficiency of the cycle is calculated using

$$\eta = \frac{W}{Q_H}$$

W is the net work of the cycle, which is the work done by the turbine, W_t, less the work input into the pump, W_p.

$$W = W_t - W_p$$

Q_H is the heat input to the system. Defining point 1 as the turbine entrance, point 2 as the turbine exit, point 3 as the pump entrance, and point 4 as the pump exit, the efficiency can be written in terms of the enthalpies.

$$\begin{aligned} \eta &= \frac{W_t - W_p}{Q_H} \\ &= \frac{(h_1 - h_2) - (h_4 - h_3)}{h_1 - h_4} \end{aligned}$$

From steam tables, for 900°F at 1000 psia,

$$h_1 = 1448.1\ \text{Btu/lbm}$$

Since expansion in the turbine is an isentropic process, h_2 is most conveniently found using a Mollier diagram.

$$h_2 = 1075\ \text{Btu/lbm}$$

The pump work is calculated by

$$W_p = h_4 - h_3 = \upsilon_f\,(p_4 - p_3)$$

υ_f is the specific volume of the saturated liquid entering the pump. From the saturated steam tables for 20 psia,

$$\begin{aligned} \upsilon_f &= 0.01683\ \text{ft}^3/\text{lbm} \\ h_3 &= 196.19\ \text{Btu/lbm} \end{aligned}$$

h_3 is needed to find h_4.

$$\begin{aligned} h_4 - h_3 &= \left(0.01683\ \frac{\text{ft}^3}{\text{lbm}}\right) \\ &\quad \times \left(1000\ \frac{\text{lbf}}{\text{in}^2} - 20\ \frac{\text{lbf}}{\text{in}^2}\right) \\ &\quad \times \left(144\ \frac{\text{in}^2}{\text{ft}^2}\right)\left(\frac{1\ \text{Btu}}{778\ \text{ft-lbf}}\right) \\ &= 3.1\ \text{Btu/lbm} \\ h_4 - 196.19\ \frac{\text{Btu}}{\text{lbm}} &= 3.1\ \text{Btu/lbm} \\ h_4 &= 199.3\ \text{Btu/lbm} \end{aligned}$$

Substituting into the efficiency equation,

$$\begin{aligned} \eta &= \frac{\left(1448.1\ \frac{\text{Btu}}{\text{lbm}} - 1075\ \frac{\text{Btu}}{\text{lbm}}\right) - 3.1\ \frac{\text{Btu}}{\text{lbm}}}{1448.1\ \frac{\text{Btu}}{\text{lbm}} - 199.3\ \frac{\text{Btu}}{\text{lbm}}} \\ &= 0.296 \quad (30\%) \end{aligned}$$

The answer is (C).

137. The enthalpy of the output mixture is found from the conservation of energy equation.

$$\sum \dot{m}_\text{entering} h_\text{entering} = \sum \dot{m}_\text{leaving} h_\text{leaving}$$

The enthalpies of the two input steams are determined using steam tables.

$$\begin{aligned} p_1 &= 200\ \text{lbf/in}^2 \\ T_1 &= 600\text{°F} \\ h_1 &= 1322.1\ \text{Btu/lbm} \end{aligned}$$

For saturated liquid,

$$\begin{aligned} p_2 &= 200\ \text{lbf/in}^2 \\ h_f &= 355.6\ \text{Btu/lbm} \\ h_{fg} &= 843.7\ \text{Btu/lbm} \end{aligned}$$

The enthalpy of the second input steam is

$$\begin{aligned} h_2 &= h_f + xh_{fg} \\ &= 355.6\ \frac{\text{Btu}}{\text{lbm}} + (0.7)\left(843.7\ \frac{\text{Btu}}{\text{lbm}}\right) \\ &= 946.2\ \text{Btu/lbm} \end{aligned}$$

The enthalpy of the output mixture can now be found.

$$\left(4\ \frac{\text{lbm}}{\text{min}}\right)\left(1322.1\ \frac{\text{Btu}}{\text{lbm}}\right) + \left(11\ \frac{\text{lbm}}{\text{min}}\right)\left(946.2\ \frac{\text{Btu}}{\text{lbm}}\right) = \left(4\ \frac{\text{lbm}}{\text{min}} + 11\ \frac{\text{lbm}}{\text{min}}\right) h_{\text{leaving}}$$

$$h_{\text{leaving}} = 1046.4\ \text{Btu/lbm}$$

The output is partially saturated vapor at 200 psi. The quality of the output mixture, x, can be found using

$$\begin{aligned} h_{\text{leaving}} &= h_f + xh_{fg} \\ 1046.4\ \frac{\text{Btu}}{\text{lbm}} &= 355.6\ \frac{\text{Btu}}{\text{lbm}} + x\left(843.7\ \frac{\text{Btu}}{\text{lbm}}\right) \\ x &= 0.819 \quad (82\%) \end{aligned}$$

The answer is (B).

138. Draw a diagram and a thermal circuit of the system.

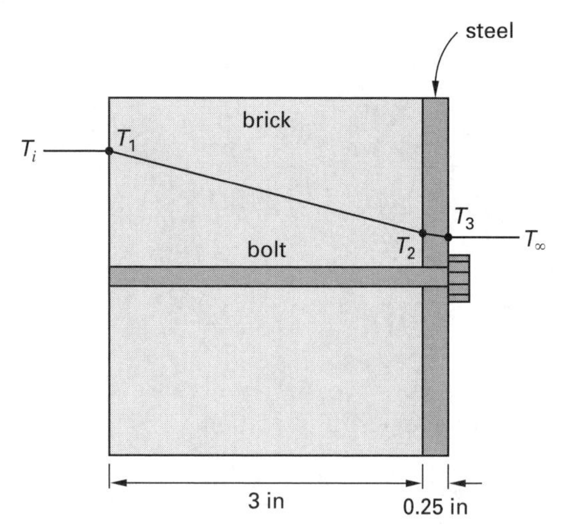

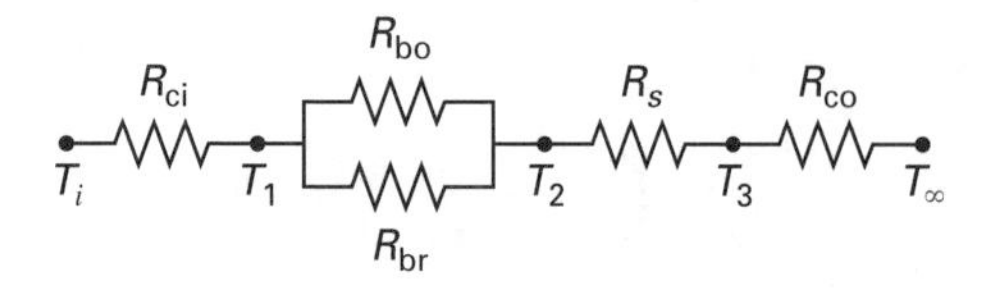

The heat flux can be found using

$$\begin{aligned} q &= \frac{\Delta T}{\Sigma R} \\ &= \frac{T_1 - T_\infty}{R_{\text{eq}} + R_s + R_{\text{co}}} \end{aligned}$$

T_1 and T_∞ are used since these are two known temperatures. The equivalent resistance of the parallel resistor combination of the bolt, R_{bo}, and the brick, R_{br}, is

$$R_{\text{eq}} = \frac{1}{\frac{1}{R_{\text{bo}}} + \frac{1}{R_{\text{br}}}}$$

Using $A = 1\ \text{ft}^2$ and a convection coefficient of $\bar{h}_{\text{co}} = 1.65\ \text{Btu/hr-ft}^2\text{-°F}$, R_{co} is

$$\begin{aligned} R_{\text{co}} &= \frac{1}{\bar{h}_{\text{co}} A} = \frac{1}{\left(1.65\ \frac{\text{Btu}}{\text{hr-ft}^2\text{-°F}}\right)(1\ \text{ft}^2)} \\ &= 0.606\ \text{hr-°F/Btu} \end{aligned}$$

The thermal conductivity of steel is

$$k_s = 26\ \text{Btu-ft/hr-ft}^2\text{-°F}$$

The thermal resistance of the steel per square foot is

$$\begin{aligned} R_s &= \frac{L_s}{Ak_s} = \frac{(0.25\ \text{in})\left(\frac{1\ \text{ft}}{12\ \text{in}}\right)}{(1\ \text{ft}^2)\left(26\ \frac{\text{Btu-ft}}{\text{hr-ft}^2\text{-°F}}\right)} \\ &= 0.0008\ \text{hr-°F/Btu} \end{aligned}$$

The thermal conductivity of fire-clay brick is

$$k_{\text{br}} = 0.58\ \text{Btu-ft/hr-ft}^2\text{-°F}$$

The area of the brick per square foot of furnace wall is

$$\begin{aligned} A_{\text{br}} &= 1\ \text{ft}^2 - A_{\text{bo}} \\ &= 1\ \text{ft}^2 - (6)\left(\frac{\pi}{4}\right)\left((0.25\ \text{in})\left(\frac{1\ \text{ft}}{12\ \text{in}}\right)\right)^2 \\ &= 0.998\ \text{ft}^2 \end{aligned}$$

$$\begin{aligned} R_{\text{br}} &= \frac{L}{Ak} = \frac{(3\ \text{in})\left(\frac{1\ \text{ft}}{12\ \text{in}}\right)}{(0.998\ \text{ft}^2)\left(0.58\ \frac{\text{Btu-ft}}{\text{hr-ft}^2\text{-°F}}\right)} \\ &= 0.432\ \text{hr-°F/Btu} \end{aligned}$$

The thermal resistance of a bolt is

$$\begin{aligned} R_{\text{bo}} &= \frac{L}{Ak} = \frac{(3\ \text{in})\left(\frac{1\ \text{ft}}{12\ \text{in}}\right)}{(0.002\ \text{ft}^2)\left(26\ \frac{\text{Btu-ft}}{\text{hr-ft}^2\text{-°F}}\right)} \\ &= 4.81\ \text{hr-°F/Btu} \end{aligned}$$

The equivalent resistance of the parallel combination is

$$R_{eq} = \frac{1}{\frac{1}{4.81 \frac{\text{hr-°F}}{\text{Btu}}} + \frac{1}{0.432 \frac{\text{hr-°F}}{\text{Btu}}}} = 0.396 \text{ hr-°F/Btu}$$

The heat flux can now be calculated.

$$q = \frac{1000°\text{F} - 70°\text{F}}{0.396 \frac{\text{hr-°F}}{\text{Btu}} + 0.0008 \frac{\text{hr-°F}}{\text{Btu}} + 0.606 \frac{\text{hr-°F}}{\text{Btu}}} = 927 \text{ Btu/hr} \quad (930 \text{ Btu/hr})$$

The answer is (D).

139. The easiest way to determine the NTU is to use heat exchanger effectiveness charts. To use the charts, the ratio of the thermal capacity rates, C_{min}/C_{max}, and the heat exchanger effectiveness, ε, must be determined.

Since all the temperatures are given, the thermal capacity rates can be found using

$$q = C\Delta T$$

For the water (the colder fluid), the thermal capacity is

$$\begin{aligned} q &= C_c(T_{c,\text{leaving}} - T_{c,\text{entering}}) \\ 110{,}000 \frac{\text{Btu}}{\text{hr}} &= C_c\,(120°\text{F} - 92°\text{F}) \\ C_c &= 3929 \text{ Btu/hr-°F} \end{aligned}$$

For the air (the hotter fluid), the thermal capacity is

$$\begin{aligned} q &= C_h(T_{h,\text{leaving}} - T_{h,\text{entering}}) \\ 110{,}000 \frac{\text{Btu}}{\text{hr}} &= C_h\,(190°\text{F} - 125°\text{F}) \\ C_h &= 1692 \text{ Btu/hr-°F} \end{aligned}$$

The thermal capacity ratio is

$$\frac{C_{min}}{C_{max}} = \frac{1692 \frac{\text{Btu}}{\text{hr-°F}}}{3929 \frac{\text{Btu}}{\text{hr-°F}}} = 0.431$$

The heat exchanger effectiveness can be found using

$$\begin{aligned} q &= \varepsilon C_{min}(T_{h,\text{entering}} - T_{c,\text{entering}}) \\ 110{,}000 \frac{\text{Btu}}{\text{hr}} &= \varepsilon \left(1690 \frac{\text{Btu}}{\text{hr-°F}}\right)(190°\text{F} - 92°\text{F}) \\ \varepsilon &= 0.664 \end{aligned}$$

Using an NTU chart for a one-shell pass and two-tube pass parallel counterflow heat exchanger, the number of transfer units is

$$\text{NTU} = 1.7$$

The answer is (A).

140. Flow rate through the pump is determined from its displacement, D, and rotational speed, n. Because the pump and motor are direct coupled, the pump's rotational speed equals the motor's. The flow rate is

$$Q = Dn = \left(1.93 \frac{\text{in}^3}{\text{rev}}\right)\left(1200 \frac{\text{rev}}{\text{min}}\right) = 2316 \text{ in}^3\text{/min}$$

The required pump power is related to the pressure and the flow rate.

$$P = \frac{pQ}{\eta_{pump}} = \frac{\left(2500 \frac{\text{lbf}}{\text{in}^2}\right)\left(2316 \frac{\text{in}^3}{\text{min}}\right)}{(0.85)\left(\frac{33{,}000 \frac{\text{ft-lbf}}{\text{min}}}{1 \text{ hp}}\right)\left(12 \frac{\text{in}}{\text{ft}}\right)} = 17.20 \text{ hp}$$

The current drawn by the motor is

$$I = \frac{P}{\sqrt{3}V\eta_{motor}(\text{pf})} = \frac{(17.20 \text{ hp})\left(746 \frac{\text{W}}{\text{hp}}\right)}{\sqrt{3}(208 \text{ V})(0.90)(0.85)} = 46.56 \text{ A} \quad (50 \text{ A})$$

The answer is (B).

141. The amount of heat lost due to convection is

$$\begin{aligned} q &= hA(T_s - T_\infty) \\ &= h(\pi d^2)(260°\text{F} - 80°\text{F}) \\ &= h(\pi)\left((2 \text{ in})\left(\frac{1 \text{ ft}}{12 \text{ in}}\right)\right)^2(260°\text{F} - 80°\text{F}) \end{aligned}$$

The convection coefficient can be found using

$$h = \frac{\text{Nu}k}{d}$$

The Nusselt number of the sphere (based on diameter) is

$$\text{Nu} = 2 + 0.45\,(\text{GrPr})^{0.25}$$

The Grashof number for a sphere (based on diameter) is

$$\begin{aligned}\text{Gr} &= \left(\frac{g\beta\rho^2}{\mu^2}\right)(d^3)(T_s - T_\infty) \\ &= \left(\frac{g\beta\rho^2}{\mu^2}\right)(2\text{ in})^3\left(\frac{1\text{ ft}}{12\text{ in}}\right)^3(260^\circ\text{F} - 80^\circ\text{F})\end{aligned}$$

The properties of air at the mean temperature of 170°F are found using tables.

$$\begin{aligned}k &= 0.0168\text{ Btu/hr-ft-}^\circ\text{F} \\ \text{Pr} &= 0.72 \\ \frac{g\beta\rho^2}{\mu^2} &= 1.123\times 10^6\ \frac{1}{\text{ft}^3\text{-}^\circ\text{F}}\end{aligned}$$

The Grashof and Nusselt numbers are

$$\begin{aligned}\text{Gr} &= \left(1.123\times 10^6\ \frac{1}{\text{ft}^3\text{-}^\circ\text{F}}\right)(2\text{ in})^3\left(\frac{1\text{ ft}}{12\text{ in}}\right)^3 \\ &\quad\times(260^\circ\text{F} - 80^\circ\text{F}) \\ &= 9.4\times 10^5 \\ \text{Nu} &= 2 + (0.45)\left((9.4\times 10^5)(0.72)\right)^{0.25} \\ &= 14.9\end{aligned}$$

The convection coefficient is

$$\begin{aligned}h &= \frac{(14.9)\left(0.0168\ \frac{\text{Btu}}{\text{hr-ft-}^\circ\text{F}}\right)}{(2\text{ in})\left(\frac{1\text{ ft}}{12\text{ in}}\right)} \\ &= 1.5\text{ Btu/hr-ft}^2\text{-}^\circ\text{F}\end{aligned}$$

The heat loss is

$$\begin{aligned}q &= \left(1.5\ \frac{\text{Btu}}{\text{hr-ft}^2\text{-}^\circ\text{F}}\right)\pi\left((2\text{ in})\left(\frac{1\text{ ft}}{12\text{ in}}\right)\right)^2 \\ &\quad\times(260^\circ\text{F} - 80^\circ\text{F})\left(\frac{1\text{ W-hr}}{3.412\text{ Btu}}\right) \\ &= 6.9\text{ W}\end{aligned}$$

Alternate Solution

The heat loss can also be found following the procedure given in the *Mechanical Engineering Reference Manual.*

$$h = \frac{kC(\text{GrPr})^n}{L}$$

The characteristic length for a sphere is the radius. So the Grashof number becomes

$$\begin{aligned}\text{Gr} &= \left(1.123\times 10^6\ \frac{1}{\text{ft}^3\text{-}^\circ\text{F}}\right)(1\text{ in})^3\left(\frac{1\text{ ft}}{12\text{ in}}\right)^3 \\ &\quad\times(260^\circ\text{F} - 80^\circ\text{F}) \\ &= 1.17\times 10^5\end{aligned}$$

Since the product GrPr is between 10^3 and 10^9,

$$\begin{aligned}C &= 0.53 \\ n &= 0.25\end{aligned}$$

The convection coefficient becomes

$$\begin{aligned}h &= \frac{\left(0.0168\ \frac{\text{Btu}}{\text{hr-ft-}^\circ\text{F}}\right)(0.53)\left((1.17\times 10^5)(0.72)\right)^{0.25}}{(1\text{ in})\left(\frac{1\text{ ft}}{12\text{ in}}\right)} \\ &= 1.82\text{ Btu/hr-ft}^2\text{-}^\circ\text{F}\end{aligned}$$

This yields a heat loss of

$$\begin{aligned}q &= \left(1.82\ \frac{\text{Btu}}{\text{hr-ft}^2\text{-}^\circ\text{F}}\right)\pi\left((2\text{ in})\left(\frac{1\text{ ft}}{12\text{ in}}\right)\right)^2 \\ &\quad\times(260^\circ\text{F} - 80^\circ\text{F})\left(\frac{1\text{ W-hr}}{3.412\text{ Btu}}\right) \\ &= 8.4\text{ W}\quad\text{(The closest option is 6.9 W.)}\end{aligned}$$

The answer is (B).

142. This is a two-dimensional, steady-state conduction problem. If the surface temperature of the insulation can be determined, then the heat loss through the ground can be found.

For a first approximation, assume the pipe is at the same temperature as the glycol solution. Maximum heat loss will occur where the pipe enters the ground, where the temperature differential between the insulation surface and the ground is greatest.

For 1 in schedule-40 pipe, the outside diameter is $d_o = 1.315$ in. The outside diameter of the wrapped pipe is $d_{o,\text{wrapped}} = 1.315\text{ in} + (2)(0.125\text{ in}) = 1.565$ in.

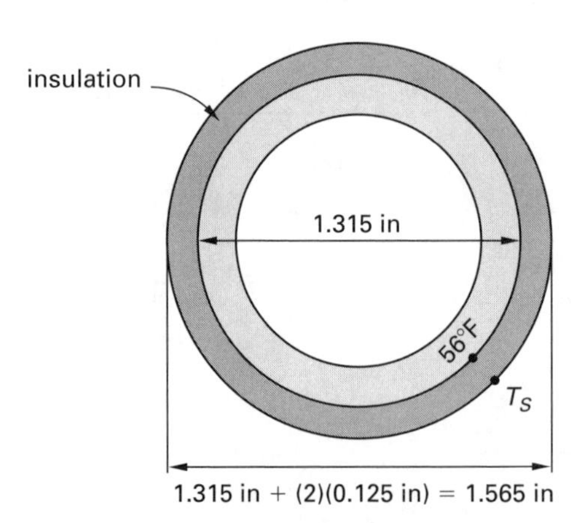

The resistance of the insulation per unit length can be found using

$$R = \frac{\ln \frac{r_o}{r_i}}{2\pi k L} = \frac{\ln\left(\frac{\frac{1.565 \text{ in}}{2}}{\frac{1.315 \text{ in}}{2}}\right)}{2\pi\left(0.03 \frac{\text{Btu}}{\text{hr-ft-°F}}\right)L}$$

$$= \frac{0.923 \frac{\text{hr-ft-°F}}{\text{Btu}}}{L}$$

The heat transfer due to conduction through the insulation is

$$q = \frac{\Delta T}{R} = \frac{56°\text{F} - T_s}{\frac{0.923 \frac{\text{hr-ft-°F}}{\text{Btu}}}{L}}$$

$$\frac{q}{L} = \left(1.08 \frac{\text{Btu}}{\text{hr-ft-°F}}\right)(56°\text{F} - T_s)$$

The heat conduction through the ground is

$$q = Sk\Delta T$$

S is the shape factor. For a cylinder underground, S is

$$S = \frac{2\pi L}{\cosh^{-1}\frac{2z}{d}}$$

$$= \frac{2\pi L}{\cosh^{-1}\left(\frac{(2)(3 \text{ ft})}{(1.565 \text{ in})\left(\frac{1 \text{ ft}}{12 \text{ in}}\right)}\right)}$$

$$= 1.39L$$

Although not specified, the soil is likely to be saturated due to the proximity to the pond. The soil thermal conductivity is taken as 1.5 Btu/hr-ft-°F. Substitute this and known values into the heat conduction equation.

$$q = (1.39L)\left(1.5 \frac{\text{Btu}}{\text{hr-ft-°F}}\right)(T_s - 20°\text{F})$$

$$\frac{q}{L} = \left(2.085 \frac{\text{Btu}}{\text{hr-ft-°F}}\right)(T_s - 20°\text{F})$$

The heat transfer rates are equal.

$$\left(1.08 \frac{\text{Btu}}{\text{hr-ft-°F}}\right)(56°\text{F} - T_s) = \left(2.085 \frac{\text{Btu}}{\text{hr-ft-°F}}\right)(T_s - 20°\text{F})$$

$$T_s = 32.3°\text{F}$$

The heat transfer rate per unit length is

$$\frac{q}{L} = \left(0.0417 \frac{\text{Btu}}{\text{hr-ft-°F}}\right)(32.3°\text{F} - 20°\text{F})$$

$$= 0.51 \text{ Btu/hr-ft} \quad (0.5 \text{ Btu/hr-ft})$$

The answer is (A).

143. Calculate the hydraulic power.

$$P_{\text{hyd}} = h\dot{m}\left(\frac{g}{g_c}\right)$$

$$= \left(\frac{(300 \text{ ft})\left(1.5 \frac{\text{lbm}}{\text{sec}}\right)}{550 \frac{\text{ft-lbf}}{\text{hp-sec}}}\right)\left(\frac{32.2 \frac{\text{ft}}{\text{sec}}}{32.2 \frac{\text{ft-lbm}}{\text{lbf-sec}^2}}\right)$$

$$= 0.818 \text{ hp}$$

Calculate the motor power.

$$P_{\text{motor}} = \frac{P_{\text{hyd}}}{\eta_{\text{motor}}\eta_{\text{pump}}} = \frac{0.818 \text{ hp}}{(0.95)(0.65)} = 1.3 \text{ hp}$$

Calculate the motor current.

$$I_{\text{motor}} = \frac{P_{\text{motor}}}{V_{\text{motor}}} = \left(\frac{1.32 \text{ hp}}{240 \text{ V}}\right)\left(746 \frac{\text{W}}{\text{hp}}\right)$$

$$= 4.1 \text{ A}$$

The answer is (B).

144. The rod will buckle when the thermal stress reaches the Euler stress. To find the Euler stress, check the slenderness ratio. The radius of gyration is

$$k = \sqrt{\frac{I}{A}} = \sqrt{\frac{\frac{\pi r^4}{4}}{\pi r^2}}$$

$$= \frac{r}{2} = \frac{1.2 \text{ in}}{2}$$

$$= 0.6 \text{ in}$$

For two built-in ends, the end restraint coefficient has a design value of $C = 0.65$. The slenderness ratio is

$$\frac{CL}{k} = \frac{(0.65)(11.5 \text{ ft})\left(12 \frac{\text{in}}{\text{ft}}\right)}{0.6 \text{ in}}$$

$$= 149.5$$

The rod may be regarded as a long column. The Euler stress for the rod is thus

$$\sigma_e = \frac{\pi^2 E}{\left(\frac{CL}{k}\right)^2}$$

The thermal stress for the rod can be expressed as a function of the temperature change.

$$\sigma_{\text{th}} = E\varepsilon_{\text{th}} = E\alpha\Delta T$$

Find the temperature change that will cause the thermal stress to equal the Euler stress, by equating the thermal and Euler stresses and solving for the temperature change.

$$\sigma_{\text{th}} = \sigma_e$$

$$E\alpha\Delta T = \frac{\pi^2 E}{\left(\frac{CL}{k}\right)^2}$$

$$\Delta T = \frac{\pi^2}{\alpha\left(\frac{CL}{k}\right)^2}$$

The average coefficient of linear thermal expansion, α, for steel is 6.5×10^{-6} 1/°F.

$$\begin{aligned}\Delta T &= \frac{\pi^2}{\alpha\left(\frac{CL}{k}\right)^2} \\ &= \frac{\pi^2}{\left(6.5 \times 10^{-6}\ \frac{1}{^\circ\text{F}}\right)(149.5)^2} \\ &= 67.94^\circ\text{F} \quad (68^\circ\text{F})\end{aligned}$$

The answer is (A).

145. First construct a thermal circuit of the system.

$1/h_cA$ $1/h_rA$

T_{flame} q_c T_{thermo} q_r T_{wall}

The heat transfer per unit area due to convection is

$$\begin{aligned}\frac{q_c}{A} &= h_c(T_{\text{flame}} - T_{\text{thermo}}) \\ &= \left(25\ \frac{\text{Btu}}{\text{hr-ft}^2\text{-}^\circ\text{F}}\right)(T_{\text{flame}} - 550^\circ\text{F})\end{aligned}$$

The heat transfer per unit area due to radiation is

$$\frac{q_r}{A} = \varepsilon\sigma F_A\left(T_{\text{thermo}}^4 - T_{\text{wall}}^4\right)$$

The maximum flame temperature will be obtained by assuming all the surfaces to be black. The emissivity, ε, will then be equal to one. Since the thermocouple is in an enclosed surface with large area, the shape factor, F_A, is also equal to one.

$$\begin{aligned}T_{^\circ\text{R}} &= T_{^\circ\text{F}} + 460^\circ \\ T_{\text{flame}} &= 550^\circ\text{F} + 460^\circ = 1010^\circ\text{R} \\ T_{\text{wall}} &= 400^\circ\text{F} + 460^\circ = 860^\circ\text{R} \\ \frac{q_r}{A} &= (1)\left(0.1714 \times 10^{-8}\frac{\text{Btu}}{\text{hr-ft}^2\text{-}^\circ\text{R}^4}\right)(1) \\ &\quad \times \left((1010^\circ\text{R})^4 - (860^\circ\text{R})^4\right) \\ &= 846\ \text{Btu/hr-ft}^2\end{aligned}$$

The heat transfer rates are equal (i.e., $q_c = q_r$).

$$\begin{aligned}846\ \frac{\text{Btu}}{\text{hr-ft}^2} &= \left(25\ \frac{\text{Btu}}{\text{hr-ft}^2\text{-}^\circ\text{F}}\right)(T_{\text{flame}} - 550^\circ\text{F}) \\ T_{\text{flame}} &= 584^\circ\text{F} \quad (580^\circ\text{F})\end{aligned}$$

The answer is (A).

146. The following diagrams apply to this problem.

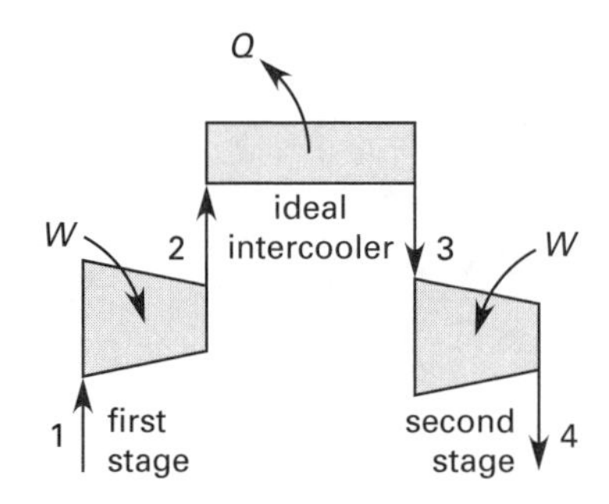

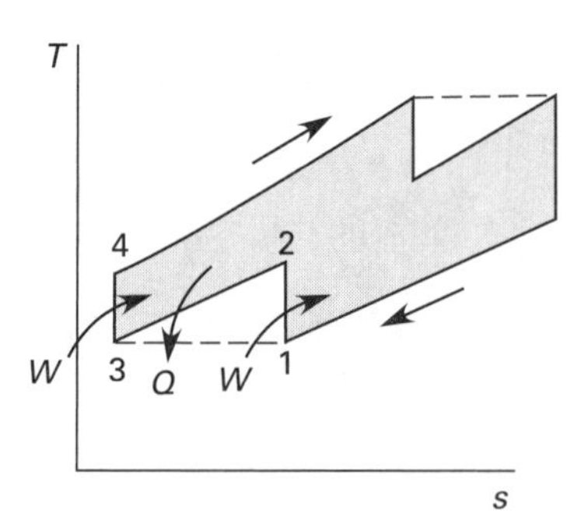

For an ideal intercooler,

$$T_1 = T_3 = 70^\circ\text{F}$$

$$p_2 = p_3$$

The heat removed in the intercooler is

$$\begin{aligned}\dot{Q} &= \dot{m}_a c_p(T_2 - T_3) \\ &= \left(10\ \frac{\text{lbm}}{\text{min}}\right)\left(0.24\ \frac{\text{Btu}}{\text{lbm-}^\circ\text{F}}\right)(T_2 - 70^\circ\text{F})\end{aligned}$$

For isentropic compressions,

$$\frac{T_2}{T_1} = \left(\frac{p_2}{p_1}\right)^{\frac{k-1}{k}}$$

For air, $k = 1.4$, and T is in degrees Rankine.

$$T_{^\circ\text{R}} = T_{^\circ\text{F}} + 460^\circ$$
$$T_1 = 70^\circ\text{F} + 460^\circ = 530^\circ\text{R}$$
$$T_2 = (530^\circ\text{R})\left(\frac{p_2}{14.7\ \frac{\text{lbf}}{\text{in}^2}}\right)^{\frac{1.4-1}{1.4}}$$

Determine p_2. For ideal staging,

$$\frac{p_2}{p_1} = \frac{p_4}{p_3}$$
$$p_3 = p_2$$
$$p_2^2 = p_1 p_4$$
$$p_2 = \sqrt{\left(14.7\ \frac{\text{lbf}}{\text{in}^2}\right)\left(200\ \frac{\text{lbf}}{\text{in}^2}\right)}$$
$$= 54.2\ \text{lbf/in}^2$$

Find T_2.

$$T_2 = (530^\circ\text{R})\left(\frac{54.2\ \frac{\text{lbf}}{\text{in}^2}}{14.7\ \frac{\text{lbf}}{\text{in}^2}}\right)^{\frac{1.4-1}{1.4}}$$
$$= 769^\circ\text{R} - 460^\circ$$
$$= 309^\circ\text{F}$$

Now, $\dot{Q}$ can be found.

$$\dot{Q} = \left(10\ \frac{\text{lbm}}{\text{min}}\right)\left(0.24\ \frac{\text{Btu}}{\text{lbm-}^\circ\text{F}}\right)(309^\circ\text{F} - 70^\circ\text{F})$$
$$= 574\ \text{Btu/min} \quad (570\ \text{Btu/min})$$

The answer is (A).

147. Consider point 1 to be at the surface of the water in the tank and point 2 to be just outside the water's exit. Bernoulli's equation is

$$\frac{p_1}{\rho} + \frac{\text{v}_1^2}{2g_c} + \frac{z_1 g}{g_c} = \frac{p_2}{\rho} + \frac{\text{v}_2^2}{2g_c} + \frac{z_2 g}{g_c}$$

Both points are at atmospheric pressure, so the p/ρ terms are equal and can be eliminated. The surface of the water in the tank is stationary, so v_1 is zero, as is the term containing it. The remaining terms give

$$\frac{z_1 g}{g_c} = \frac{\text{v}_2^2}{2g_c} + \frac{z_2 g}{g_c}$$

Solving for the velocity at the exit,

$$\text{v}_2 = \sqrt{2g(z_1 - z_2)}$$
$$= \sqrt{(2)\left(32.2\ \frac{\text{ft}}{\text{sec}^2}\right)(11\ \text{ft})}$$
$$= 26.62\ \text{ft/sec} \quad (27\ \text{ft/sec})$$

The answer is (C).

148. The following illustration shows the cooling-reheat process on a psychrometric chart. To remove moisture from the air, it is first cooled from state 1 to state 2′ at a constant humidity ratio, ω, so that the air becomes saturated ($\phi = 100\%$). At point 2′ the water vapor begins to condense, and further cooling to state 2 reduces the humidity ratio. The air is then heated to the desired temperature at the humidity ratio of state 2. This final point is state 3.

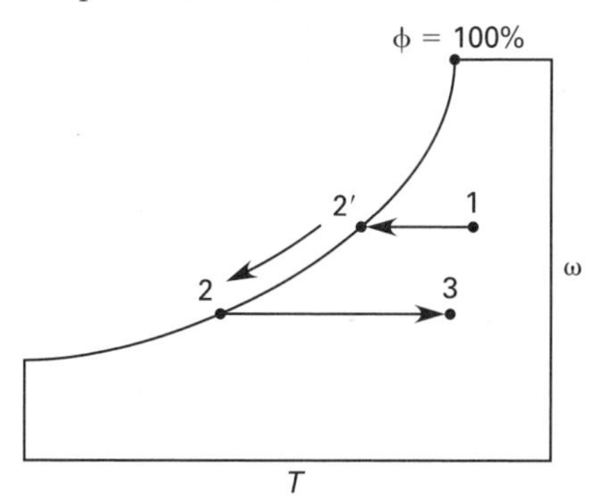

The heat removed during the cooling process is given by

$$\dot{q}_r = \dot{m}_a\,(h_1 - h_2 - (\omega_1 - \omega_2)h_f)$$

The term $(\omega_1 - \omega_2)h_f$ is small compared to other enthalpy terms and is often neglected.

Since there is no water lost or gained in the heating section of the evaporator, the heat added during the heating process is given by

$$\dot{q}_a = \dot{m}_a\,(h_3 - h_2)$$

The enthalpies h_1, h_2, and h_3 can be found using a psychrometric chart. First, locate point 1 where the dry-bulb temperature is 98°F, and then find the diagonal line sloping down from left to right for the 75°F wet-bulb temperature. Then, follow this diagonal line up and to the left to find $h_1 = 38.6$ Btu/lbm.

Second, locate point 3. This is the point where the dry-bulb temperature is 67°F and the relative humidity is 50%. Follow the diagonal line that passes through this point up and to the left to find $h_3 = 23.8$ Btu/lbm. To locate point 2, find point 3 again and then travel left on a horizontal line of constant specific humidity until the 100% humidity line is reached (the leftmost upward curving line). The intersection of the horizontal line and the 100% humidity line is point 2. Follow the diagonal line that passes through this point up and to the left to find $h_2 = 19.2$ Btu/lbm.

The air mass flow rate can be found from the volumetric flow rate and the specific volume, v, of the dry air at the state 1 (inlet) conditions. The specific volume of dry air can be found using the psychrometric chart.

$$\dot{m}_a = \frac{\dot{V}}{v_{a,1}} = \frac{800\ \frac{\text{ft}^3}{\text{min}}}{14.37\ \frac{\text{ft}^3}{\text{lbm}}} = 55.7\ \text{lbm/min}$$

Substitute the air mass flow rate to find the rate at which heat is being removed.

$$q_r = \left(55.7\ \frac{\text{lbm}}{\text{min}}\right)\left(60\ \frac{\text{min}}{\text{hr}}\right)\left(38.6\ \frac{\text{Btu}}{\text{lbm}} - 19.2\ \frac{\text{Btu}}{\text{lbm}}\right) = 64{,}835\ \text{Btu/hr}$$

Substitute the air mass flow rate to find the rate of heat addition.

$$q_a = \left(55.7\ \frac{\text{lbm}}{\text{min}}\right)\left(60\ \frac{\text{min}}{\text{hr}}\right)\left(23.8\ \frac{\text{Btu}}{\text{lbm}} - 19.2\ \frac{\text{Btu}}{\text{lbm}}\right) = 15{,}373\ \text{Btu/hr}$$

The answer is (D).

149. The following diagrams apply to this problem.

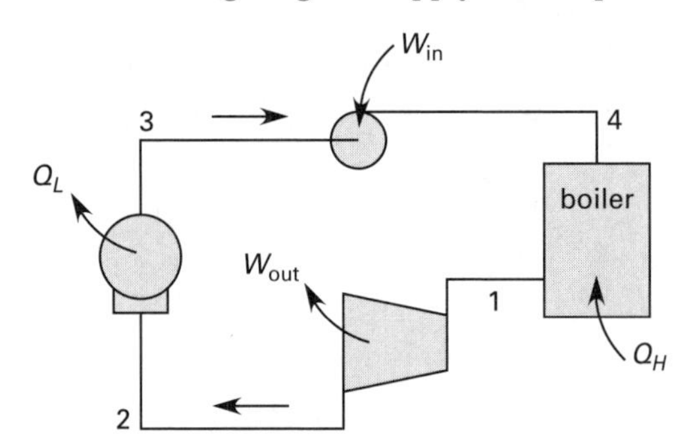

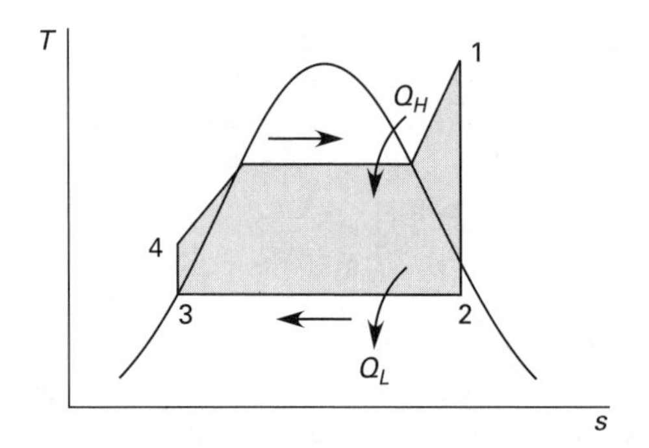

State 2 is at the output of the turbine where the temperature and steam quality are known. Since turbine expansion is an isentropic process, once the entropy at state 2 is found, the entropy at state 1 is also known. With this and the given pressure, the temperature of the steam entering the turbine can be found.

The entropy at state 2 can be found using

$$s_2 = s_f + x s_{fg}$$

Using the saturated steam tables at 100° F,

$$s_2 = 0.12963\ \frac{\text{Btu}}{\text{lbm-°R}} + (0.92)\left(1.8526\ \frac{\text{Btu}}{\text{lbm-°R}}\right) = 1.834\ \text{Btu/lbm-°R}$$

Since $s_1 = s_2$ and the pressure is 250 psia, using the superheated steam tables, the entering steam temperature is between 1040°F and 1060°F. This temperature can be found by linear interpolation between the 1040°F and 1060°F temperatures.

$$\frac{T_{1060°F} - T_{1040°F}}{s_{1060°F} - s_{1040°F}} = \frac{T_1 - T_{1040°F}}{s_1 - s_{1040°F}}$$

$$\frac{1060°F - 1040°F}{\left(1.8386\ \frac{\text{Btu}}{\text{lbm-°R}} - 1.8315\ \frac{\text{Btu}}{\text{lbm-°R}}\right)} = \frac{T_1 - 1040°F}{\left(1.834\ \frac{\text{Btu}}{\text{lbm-°R}} - 1.8315\ \frac{\text{Btu}}{\text{lbm-°R}}\right)}$$

$$T_1 = 1047°F\quad (1000°F)$$

The answer is (C).

150. The T-s diagram for an air-standard Otto cycle is shown.

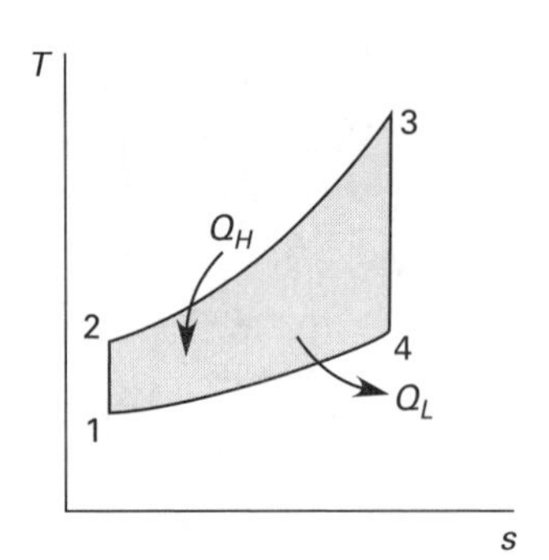

The thermal efficiency of an Otto cycle can be computed using

$$\eta_{th} = 1 - r^{1-k}$$

For air $k = 1.4$.

$$\eta_{th} = 1 - (8.5)^{1-1.4} = 0.575 \quad (57.5\%)$$

The efficiency of a Carnot cycle is given by

$$\begin{aligned}\eta_{th} &= 1 - \frac{T_L}{T_H} \\ T_{\circ R} &= T_{\circ F} + 460^\circ \\ T_1 &= 68^\circ F + 460^\circ \\ &= 528^\circ R \\ \eta_{th} &= 1 - \frac{528^\circ R}{T_H}\end{aligned}$$

T_H, the hottest temperature in the cycle, occurs at point 3, along with the highest pressure.

$$\begin{aligned}\frac{T_3}{T_2} &= \frac{p_3}{p_2} \\ T_3 &= T_2\left(\frac{510\ \frac{\text{lbf}}{\text{in}^2}}{p_2}\right)\end{aligned}$$

The temperature T_2 can be calculated using

$$\frac{T_2}{T_1} = \left(\frac{V_1}{V_2}\right)^{k-1}$$

V_1/V_2 is equal to the compression ratio.

$$T_2 = (528^\circ R)(8.5)^{1.4-1} = 1243^\circ R$$

Find the pressure p_2.

$$\begin{aligned}\frac{p_1V_1}{T_1} &= \frac{p_2V_2}{T_2} \\ p_2 &= \left(\frac{T_2p_1}{T_1}\right)\left(\frac{V_1}{V_2}\right) \\ &= \frac{(1243^\circ R)\left(14.7\ \frac{\text{lbf}}{\text{in}^2}\right)}{528^\circ R}(8.5) \\ &= 294\ \text{lbf/in}^2\end{aligned}$$

Substitute for p_2, and solve for T_3.

$$\begin{aligned}T_3 &= (1243^\circ R)\left(\frac{510\ \frac{\text{lbf}}{\text{in}^2}}{294\ \frac{\text{lbf}}{\text{in}^2}}\right) \\ &= 2156^\circ R\end{aligned}$$

The Carnot efficiency is

$$\begin{aligned}\eta_{th} &= 1 - \frac{528^\circ R}{2156^\circ R} \\ &= 0.755 \quad (75.5\%)\end{aligned}$$

The difference in efficiency is

$$\begin{aligned}\Delta\eta &= 75.5\% - 57.5\% \\ &= 18\% \quad (20\%)\end{aligned}$$

The answer is (B).

151. The following illustrations apply to this problem.

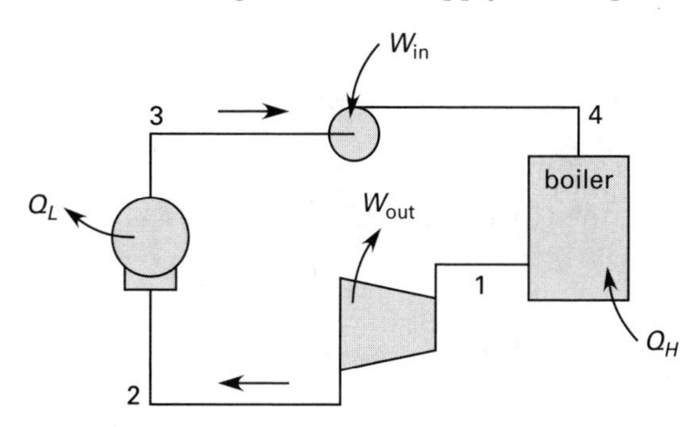

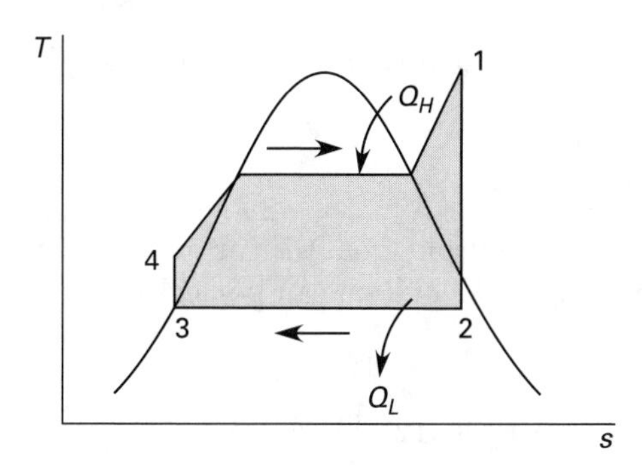

The mass flow rate can be determined using the net power output from the system.

$$\begin{aligned}P_{net} &= P_{turb} - P_{pump} \\ &= \dot{m}(h_1 - h_2) - \dot{m}(h_4 - h_3)\end{aligned}$$

The enthalpy at state 1 (1100°F and 1250 psia) is found in the superheated steam tables.

$$h_1 = 1556.6\ \text{Btu/lbm}$$

The enthalpy at state 2 is found using a Mollier diagram. Since the turbine is an isentropic process, locate state 1 on the diagram and drop straight down to the 5 psia curve. The enthalpy is

$$h_2 = 1018\ \text{Btu/lbm}$$

State 3 is saturated liquid at 5 psia.

$$h_3 = 130.17 \text{ Btu/lbm}$$

State 4 is compressed liquid. The enthalpy can be computed using

$$\begin{aligned} h_4 &= h_3 + \upsilon_3(p_4 - p_3) \\ &= 130.17 \frac{\text{Btu}}{\text{lbm}} + \left(0.016407 \frac{\text{ft}^3}{\text{lbm}}\right) \\ &\quad \times \left(1250 \frac{\text{lbf}}{\text{in}^2} - 5 \frac{\text{lbf}}{\text{in}^2}\right)\left(144 \frac{\text{in}^2}{\text{ft}^2}\right) \\ &\quad \times \left(\frac{1 \text{ Btu}}{778 \text{ ft-lbf}}\right) \\ &= 133.95 \text{ Btu/lbm} \end{aligned}$$

Substitute and solve for the mass flow rate.

$$\begin{aligned} &(250 \times 10^6 \text{ W})\left(3.412 \frac{\frac{\text{Btu}}{\text{hr}}}{\text{W}}\right) \\ &\quad = \dot{m}\left(1556.6 \frac{\text{Btu}}{\text{lbm}} - 1018 \frac{\text{Btu}}{\text{lbm}}\right) \\ &\quad\quad - \dot{m}\left(133.95 \frac{\text{Btu}}{\text{lbm}} - 130.17 \frac{\text{Btu}}{\text{lbm}}\right) \\ &\dot{m} = 1.59 \times 10^6 \text{ lbm/hr} \quad (1.6 \times 10^6 \text{ lbm/hr}) \end{aligned}$$

Alternate Solution

Since the power output from the turbine is so large, the pump work into the system can be neglected.

$$\begin{aligned} &(250 \times 10^6 \text{ W})\left(3.412 \frac{\frac{\text{Btu}}{\text{hr}}}{\text{W}}\right) \\ &\quad = \dot{m}\left(1556.6 \frac{\text{Btu}}{\text{lbm}} - 1018 \frac{\text{Btu}}{\text{lbm}}\right) \\ &\dot{m} = 1.58 \times 10^6 \text{ lbm/hr} \quad (1.6 \times 10^6 \text{ lbm/hr}) \end{aligned}$$

The answer is (D).

152. The total number of spring coils includes both the active coils and the end coils. For squared and ground ends, the number of active coils is

$$n_{\text{active}} = n_{\text{total}} - 2$$

The force on the spring valve is a result of the pressure acting on the control area.

$$F = pA$$

The maximum spring deflection is

$$\delta = 7/16 \text{ in} = 0.4375 \text{ in}$$

The load deflection equation relates the properties of the spring to the applied force.

$$\begin{aligned} \frac{F}{\delta} &= \left(\frac{Gd}{8C^3}\right)\left(\frac{1}{n_{\text{active}}}\right) \\ n_{\text{active}} &= \frac{Gd\delta}{8C^3F} = \frac{Gd\delta}{8C^3(pA)} \\ &= \frac{\left(11.5 \times 10^6 \frac{\text{lbf}}{\text{in}^2}\right)(0.10 \text{ in})(0.4375 \text{ in})}{(8)(8)^3\left(250 \frac{\text{lbf}}{\text{in}^2}\right)(0.123 \text{ in}^2)} \\ &= 3.995 \\ n_{\text{total}} &= n_{\text{active}} + 2 \\ &= 3.995 + 2 \\ &= 5.995 \quad (6 \text{ coils}) \end{aligned}$$

The answer is (C).

153. The cycle and T-s diagrams are shown.

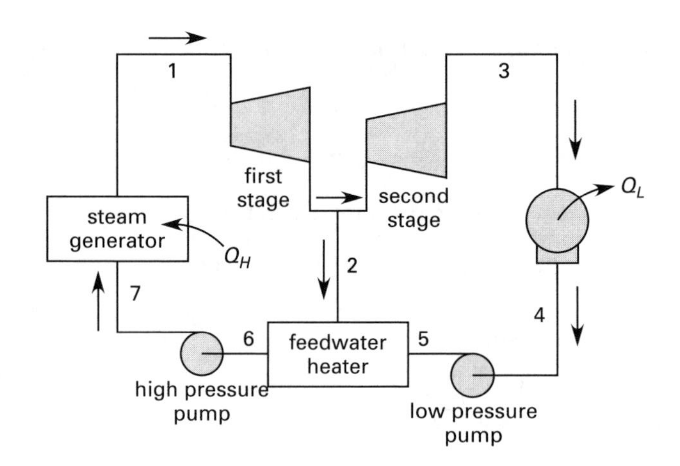

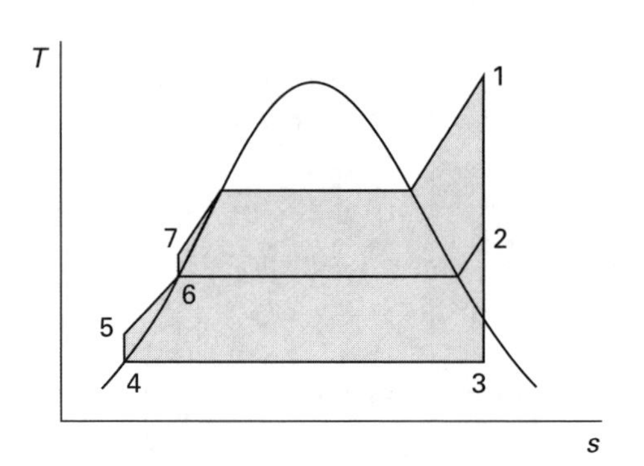

The fraction of the total flow passing through the second-stage turbine is given by

$$\frac{\dot{m}_3}{\dot{m}_1} = 1 - \frac{\dot{m}_2}{\dot{m}_1} = 1 - y$$

The fraction of total flow diverted to the open feedwater heater is y. The fraction y can be found from an energy balance for a control volume surrounding the feedwater heater.

$$y = \frac{h_6 - h_5}{h_2 - h_5}$$

Since $s_1 = s_2 = s_3$, the enthalpies at these states can be found from a Mollier diagram. h_1 and h_3 are not needed for this problem.

$$h_1 = 1440 \text{ Btu/lbm} \quad [860^\circ\text{F, 600 psia}]$$
$$h_2 = 1260 \text{ Btu/lbm} \quad [130 \text{ psia}]$$
$$h_3 = 1060 \text{ Btu/lbm} \quad [10 \text{ psia}]$$

State 6 is saturated liquid. For an ideal open feedwater heater, $p_2 = p_5 = p_6 = 130$ psia. From the saturated liquid table,

$$h_6 = 319 \text{ Btu/lbm}$$

The enthalpy at state 4 is known since the water is saturated liquid at 10 psia. The enthalpy at state 5 can be found using the pump work.

$$h_5 - h_4 = v_f(p_5 - p_4)$$

From the saturated liquid table,

$$h_4 = 161.23 \text{ Btu/lbm}$$
$$v_4 = 0.01659 \text{ ft}^3\text{/lbm}$$

$$h_5 - 161.23 \ \frac{\text{Btu}}{\text{lbm}} = \left(0.01659 \ \frac{\text{ft}^3}{\text{lbm}}\right) \times \left(130 \ \frac{\text{lbf}}{\text{in}^2} - 10 \ \frac{\text{lbf}}{\text{in}^2}\right) \times \left(144 \ \frac{\text{in}^2}{\text{ft}^2}\right)\left(\frac{1 \text{ Btu}}{778 \text{ ft-lbf}}\right)$$

$$h_5 = 161.6 \text{ Btu/lbm}$$

The fraction flowing through the second turbine is

$$\frac{\dot{m}_3}{\dot{m}_1} = 1 - \frac{319 \ \frac{\text{Btu}}{\text{lbm}} - 161.6 \ \frac{\text{Btu}}{\text{lbm}}}{1260 \ \frac{\text{Btu}}{\text{lbm}} - 161.6 \ \frac{\text{Btu}}{\text{lbm}}} = 0.857 \quad (86\%)$$

The answer is (D).

154. The specific heat for copper is 0.094 Btu/lbm-°F. The specific heat for stainless steel is 0.11 Btu/lbm-°F. The final temperature is found from the equation

$$(mc_p\Delta T)_{\text{Cu}} = (mc_p\Delta T)_{\text{st}}$$
$$(mc_p)_{\text{Cu}}(T_f - T_{i,\text{Cu}}) = (mc_p)_{\text{st}}(T_{i,\text{st}} - T_f)$$

$$(60 \text{ lbm})\left(0.094 \ \frac{\text{Btu}}{\text{lbm-}^\circ\text{F}}\right)(T_f - 50^\circ\text{F}) = (30 \text{ lbm})\left(0.11 \ \frac{\text{Btu}}{\text{lbm-}^\circ\text{F}}\right)(420^\circ\text{F} - T_f)$$

$$T_f = 186.6^\circ\text{F} \quad (190^\circ\text{F})$$

The answer is (B).

155. Location Class 1 applies to any one-mile section containing 10 or fewer buildings intended for human occupancy. Typical areas in this class include wetlands, deserts, mountains, grazing land, farmland, and sparsely populated areas.

Location Class 2 applies to any one-mile section containing more than 10 but fewer than 46 buildings intended for human occupancy. Typical areas in this class include fringe areas around cities and towns, industrial areas, ranches, and country estates.

Location Class 3 applies to any one-mile section containing 46 or more buildings intended for human occupancy, except when Location Class 4 applies. Typical areas in this class include suburban housing developments, shopping centers, and residential areas.

Location Class 4 applies to areas where most buildings are four stories and taller, where traffic is heavy or dense, and where a number of other utilities may be underground.

With 25 homes in the neighborhood, and none over two stories, Location Class 2 best applies to this area.

The answer is (B).

156. Convert the flow rate to cubic feet per second.

$$\dot{V} = \left(1800 \ \frac{\text{gal}}{\text{min}}\right)\left(\frac{1 \text{ ft}^3}{7.48 \text{ gal}}\right)\left(\frac{1 \text{ min}}{60 \text{ sec}}\right) = 4.01 \text{ ft}^3\text{/sec}$$

Convert the diameter to feet.

$$d = (10 \text{ in})\left(\frac{1 \text{ ft}}{12 \text{ in}}\right) = 0.833 \text{ ft}$$

The velocity is

$$v = \frac{\dot{V}}{A} = \frac{\dot{V}}{\pi\left(\frac{d}{2}\right)^2} = \frac{4.01\ \frac{\text{ft}^3}{\text{sec}}}{\pi\left(\frac{0.833\ \text{ft}}{2}\right)^2}$$
$$= 7.36\ \text{ft/sec}$$

The head loss is

$$h_f = \frac{fLv^2}{2dg} = \frac{(0.0195)(185\ \text{ft})\left(7.36\ \frac{\text{ft}}{\text{sec}}\right)^2}{(2)(0.833\ \text{ft})\left(32.2\ \frac{\text{ft}}{\text{sec}^2}\right)}$$
$$= 3.64\ \text{ft}$$

Convert the head loss to pressure drop.

$$\Delta p = h_f\rho\left(\frac{g}{g_c}\right)$$
$$= (3.64\ \text{ft})\left(62.4\ \frac{\text{lbm}}{\text{ft}^3}\right)\left(\frac{32.2\ \frac{\text{ft}}{\text{sec}^2}}{32.2\ \frac{\text{lbm-ft}}{\text{lbf-sec}^2}}\right) \times \left(\frac{1\ \text{ft}^2}{144\ \text{in}^2}\right)$$
$$= 1.58\ \text{lbm/in}^2 \quad (1.6\ \text{psi})$$

The answer is (A).

157. The general combustion reaction for a hydrocarbon, C_nH_m, burning with theoretical air is

$$C_nH_m + xO_2 + 3.76xN_2 \rightarrow cCO_2 + dH_2O + 3.76xN_2$$

The solution is based on 1 kmol of fuel.

$$0.8032CH_4 + 0.0575C_2H_6 + 0.0179C_3H_8 + 0.0169C_4H_{10} + 0.1045N_2 + xO_2 + 3.76xN_2 \rightarrow c\left(\begin{array}{c}0.079CO_2 + 0.001CO \\ +0.070O_2 + 0.85N_2\end{array}\right) + dH_2O$$

The nitrogen ($3.76x$) has been grouped with the products.

Balance the carbon.

$$0.8032 + (2)(0.0575) + (3)(0.0179) + (4)(0.0169) = c(0.079 + 0.001)$$
$$c = 12.994$$

Balance the hydrogen.

$$(4)(0.8032) + (6)(0.0575) + (8)(0.0179) + (10)(0.0169) = 2d$$
$$d = 1.935$$

Balance the oxygen. (The variable x could also be found from a nitrogen balance.)

$$2x = (12.994)((2)(0.079) + 0.001 + (2)(0.07)) + 1.935$$
$$x = 2.91$$

The balanced equation is

$$0.8032CH_4 + 0.0575C_2H_6 + 0.0179C_3H_8 + 0.01690C_4H_{10} + 0.1045N_2 + 2.91O_2 + 10.94N_2 \rightarrow (12.994)\left(\begin{array}{c}0.079CO_2 + 0.001CO \\ +0.070O_2 + 0.85N_2\end{array}\right) + 1.935H_2O$$

The air/fuel ratio on a molar basis is

$$R_{\text{air/fuel}} = \frac{n_{\text{air}}}{n_{\text{fuel}}} = \frac{n_{O_2} + n_{N_2}}{n_{\text{fuel}}} = \frac{2.91\ \text{kmol air} + 10.94\ \text{kmol air}}{1\ \text{kmol fuel}}$$
$$= 13.85\ \text{kmol air/kmol fuel} \quad (14\ \text{kmol air/kmol fuel})$$

The answer is (D).

158. The PLAN formula can be used to determine the horsepower in an internal combustion engine operating on the Otto cycle.

$$P = pLAN$$

The number of power strokes per second is

$$N = \frac{2n\,(\text{no. cylinders})}{\text{no. strokes per cycle}} = \frac{\left(2\ \frac{\text{strokes}}{\text{rev}}\right)\left(4200\ \frac{\text{rev}}{\text{min}}\right)\left(\frac{1\ \text{min}}{60\ \text{sec}}\right)(6\ \text{cylinders})}{4\ \text{strokes per power stroke}}$$
$$= 210\ \text{power strokes/sec}$$

The power developed is

$$P = \left(120\ \frac{\text{lbf}}{\text{in}^2}\right)\left(144\ \frac{\text{in}^2}{\text{ft}^2}\right)(84.4\ \text{mm})\left(0.00328\ \frac{\text{ft}}{\text{mm}}\right) \times \left(\frac{\pi}{4}\right)\left((100.4\ \text{mm})\left(0.00328\ \frac{\text{ft}}{\text{mm}}\right)\right)^2 \times \left(210\ \frac{\text{power strokes}}{\text{sec}}\right)\left(\frac{1\ \text{hp}}{550\ \frac{\text{ft-lbf}}{\text{sec}}}\right)$$
$$= 155.6\ \text{hp} \quad (160\ \text{hp})$$

The answer is (B).

159. The energy lost can be determined from the energy balance for a reacting system.

$$\dot{Q}_{CV} = P_{CV} + \dot{n}_F(\bar{h}_P - \bar{h}_R)$$

The power generated is

$$P_{CV} = (110 \text{ hp})\left(2545 \frac{\frac{\text{Btu}}{\text{hr}}}{\text{hp}}\right)\left(\frac{1 \text{ hr}}{3600 \text{ sec}}\right) = 77.76 \text{ Btu/sec}$$

The molar fuel flow rate is

$$\dot{n}_F = \frac{\dot{m}}{\text{MW}} = \frac{0.01 \frac{\text{lbm}}{\text{sec}}}{44 \frac{\text{lbm}}{\text{lbmol}}} = 2.27 \times 10^{-4} \text{ lbmol/sec}$$

The enthalpy of the reactants is simply the enthalpy of formation of the propane.

$$\bar{h}_R = (\bar{h}_f)_{C_3H_8} = -44{,}680 \text{ Btu/lbmol}$$

The enthalpy of the products is more complex and is given by

$$\bar{h}_P = a(\bar{h}_f + \Delta\bar{h})_{CO_2} + b(\bar{h}_f + \Delta\bar{h})_{H_2O(g)} + (3.76x)(\bar{h}_f + \Delta\bar{h})_{N_2}$$

The variables a, b, and x represent the quantity of products created. This must be determined by balancing the combustion equation. The general combustion reaction for a hydrocarbon, C_nH_m, burning with theoretical air is

$$C_nH_m + xO_2 + 3.76xN_2 \rightarrow aCO_2 + bH_2O + 3.76xN_2$$

The values for a, b, and x come from balancing the above equation and are

$$a = n$$
$$2b = m$$
$$x = n + \frac{b}{2} = n + \frac{m}{4}$$

For propane, $n = 3$ and $m = 8$.

$$a = 3$$
$$b = 4$$
$$x = 5$$

The balanced combustion equation is

$$C_3H_8 + 5O_2 + 18.8N_2 \rightarrow 3CO_2 + 4H_2O + 18.8N_2$$

The enthalpy of the products can be found.

$$T_{°R} = T_{°F} + 460°$$
$$T_{air} = 77°F + 460° = 537°R$$
$$T_{air/fuel} = 1240°F + 460° = 1700°R$$

The enthalpy of the products is

$$\bar{h}_P = (3)\left(-169{,}300 \frac{\text{Btu}}{\text{lbmol}} + \left(17{,}101.4 \frac{\text{Btu}}{\text{lbmol}} - 4030.2 \frac{\text{Btu}}{\text{lbmol}}\right)\right) + (4)\left(-104{,}040 \frac{\text{Btu}}{\text{lbmol}} + \left(14{,}455.4 \frac{\text{Btu}}{\text{lbmol}} - 4258.3 \frac{\text{Btu}}{\text{lbmol}}\right)\right) + (18.8)\left(0 \frac{\text{Btu}}{\text{lbmol}} + \left(12{,}178.9 \frac{\text{Btu}}{\text{lbmol}} - 3729.5 \frac{\text{Btu}}{\text{lbmol}}\right)\right) = -685{,}209 \text{ Btu/lbmol}$$

The enthalpies of formation for the oxygen and nitrogen terms and the $\Delta\bar{h}$ terms for each of the reactants are zero because the air and fuel enter at 77°F.

The energy loss can now be found.

$$\dot{Q}_{CV} = 77.6 \frac{\text{Btu}}{\text{sec}} + \left(2.27 \times 10^{-4} \frac{\text{lbmol}}{\text{sec}}\right) \times \left(-685{,}209 \frac{\text{Btu}}{\text{lbmol}} - \left(-44{,}680 \frac{\text{Btu}}{\text{lbmol}}\right)\right) = -67.8 \text{ Btu/sec} \quad (-68 \text{ Btu/sec})$$

The answer is (C).

160. The saturation temperature of the water vapor pressure is the dew point of the products. The water vapor pressure is a fraction of the total pressure of the products.

$$p_{water} = x_{water}p_{total} = x_{water}\left(14.7 \frac{\text{lbf}}{\text{in}^2}\right)$$

The fraction of water vapor present can be obtained from the combustion equation. The general combustion reaction for a hydrocarbon, C_nH_m, burning with theoretical air is

$$C_nH_m + xO_2 + 3.76xN_2 \rightarrow aCO_2 + bH_2O + 3.76xN_2$$

In this case,

$$\begin{aligned} a &= n \quad \text{[from carbon balance]} \\ 2b &= m \quad \text{[from hydrogen balance]} \\ x &= n + \frac{m}{4} \quad \text{[from oxygen balance]} \end{aligned}$$

The composition of octane, C_8H_{18}, is $n = 8$ and $m = 18$, so

$$\begin{aligned} a &= 8 \\ b &= 9 \\ x &= 12.5 \end{aligned}$$

The combustion reaction, then, for theoretical air is given by

$$\begin{aligned} &C_8H_{18} + 12.5O_2 + (3.76)(12.5)N_2 \\ &\quad \rightarrow 8CO_2 + 9H_2O + (3.76)(12.5)N_2 \end{aligned}$$

The total number of moles of product is

$$\begin{aligned} N &= 8\ \frac{\text{mol}}{\text{mol-fuel}} + 9\ \frac{\text{mol}}{\text{mol-fuel}} + 47\ \frac{\text{mol}}{\text{mol-fuel}} \\ &= 64\ \text{mol/mol-fuel} \end{aligned}$$

The mole fraction of water is

$$\begin{aligned} x_{\text{water}} &= \frac{b}{N} = \frac{9\ \dfrac{\text{mol}}{\text{mol-fuel}}}{64\ \dfrac{\text{mol}}{\text{mol-fuel}}} \\ &= 0.141 \end{aligned}$$

The water vapor pressure is

$$\begin{aligned} p_{\text{water}} &= (0.141)\left(14.7\ \frac{\text{lbf}}{\text{in}^2}\right) \\ &= 2.07\ \text{lbf/in}^2 \end{aligned}$$

From the steam tables, the saturation temperature corresponding to this pressure is approximately

$$T_{\text{dew}} = 127^\circ\text{F} \quad (130^\circ\text{F}) \quad \text{[interpolated]}$$

The answer is (A).